INTERIBERICA, S.A. DE EDICIONES

GREEN WORLDS
Plants and Forest Life

Doubleday and Company Inc.,
Garden City, New York, 1977
A Windfall Book

© 1975 Interiberica, S. A. - Madrid
© 1975 Aldus Books Limited, London
SBN: 385 11340 4
Library of Congress Catalog Card No: 75 13111

Also published in parts as World of Plants and
Forest Life

GREEN WORLDS

Part 1
World of Plants

by David Bellamy

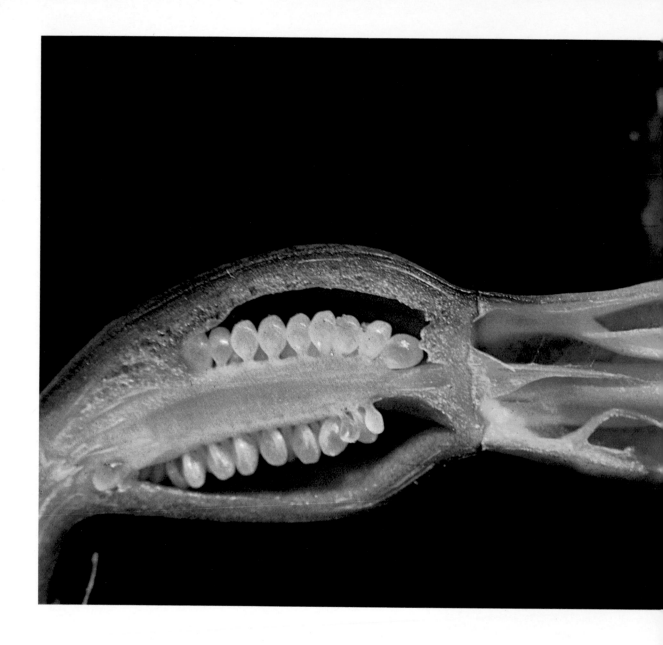

ISBN: 84-382-0014-1. Dep. Legal: M. 21.866-1975.

Printed and bound in Spain by Novograph
S.A., and Roner S.A., Crta de Irun, Km.12,450,
Madrid–34.

Series Coordinator	Geoffrey Rogers
Series Art Director	Frank Fry
Design Consultant	Guenther Radtke
Editorial Consultant	Donald Berwick
Series Consultant	Malcolm Ross-Macdonald
Art Editor	Susan Cook
Editors	Bridget Gibbs
	Damian Grint
	Maureen Cartwright
Research	Enid Moore

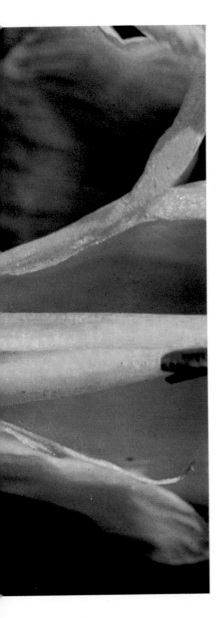

Contents: Part 1

Editorial Advisers

DAVID ATTENBOROUGH Naturalist and Broadcaster.

MICHAEL BOORER, B.SC. Author, Lecturer, and Broadcaster.

MATTHEW BRENNAN, ED.D. Director, Brentree Environmental Center, Professor of Conservation Education, Pennsylvania State University.

PHYLLIS BUSCH, ED.D. Author, Science Teacher, and Consultant in Environmental Education.

ANTHONY HUXLEY, M.A. Author and Editorial Consultant.

JAMES OLIVER, PH.D. Director of the New York Aquarium, former Director of the American Museum of Natural History, former Director of the New York Zoological Park, formerly Professor of Zoology, University of Florida.

Foreword by David Attenborough

*H*ow dramatically our view of the world would change if we could speed time. The time-lapse camera gives a hint of what we might see. With it we can watch the shoots of plant bulbs stab through the earth, unfurl their leaves, and twist their flowers around to follow the sun as it travels across the sky. But condense time on a grander scale—shorten a year into a day, encompass a forest in a single view—and the lives of plants would be seen to be titanic dramas. Speed up months and we could witness the savage battles waged in a jungle as writhing plants scramble over one another, strangling their nearest rivals in a desperate race to reach the sun. Condense a year and we could observe the springing, burgeoning and browning of a temperate woodland as one continuous episode and see the near death of winter eventually reprieved by the resurrection of spring. Lengthen the view to decades and we could watch a lake become a bog, a fen, and finally a wood; and lengthened over centuries, marvel at the way a bare rock landscape is colonized first with lichens and mosses and finally turned into a pine forest. Maybe we will never get the technology we need to produce such visions, but the botanist can see such events in his mind's eye. David Bellamy makes them come vividly alive as he charts the history and achievements of plants, from the earliest single-celled green organisms floating in water, to the highly-evolved flowering plants.

Plants were the first colonizers of the earth. From some points of view they can still be counted as the most successful. The tallest *living organism and the longest-lived are plants. Plant seeds can lie dormant for centuries and still germinate when the right conditions arrive. Plants can live in conditions where no animal could survive.*

They are also essential to animals. Without them all creatures would starve to death, for animals either eat them directly or consume other animals that have done so. And without plants, many creatures would be homeless. Michael Boorer describes the most spectacular of plant assemblies, forests. Next to coral reefs, these must be the places where life proliferates in the greatest variety. It is not easy to see their abounding richness. You can walk through a tropical rain forest for hours and fail to catch sight of a single large animal. But sit down and keep quiet and still. Within a few minutes, the forest inhabitants begin to emerge. A squirrel scampers down a tree trunk; a monkey that has been sitting motionless while there was noise from an intruder below, plucks up courage and starts feeding again on hanging fruit. A bird rockets soundlessly past, settles on the ground and begins to hunt for insects. It is only then you begin to see the wonders that are described and pictured so vividly in the following pages.

Green for Go

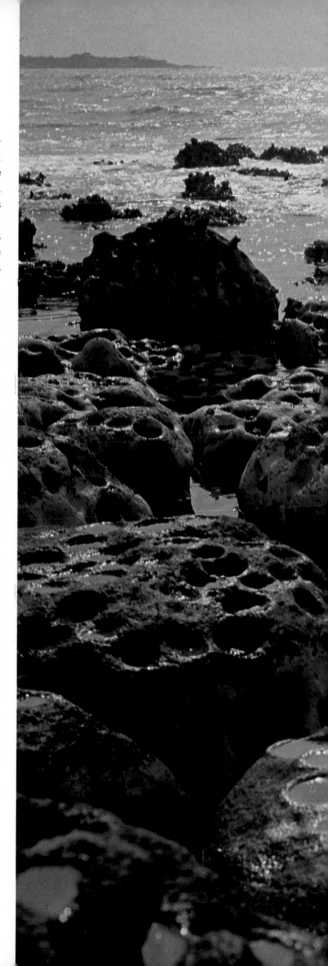

Early philosophers thought that the world consisted of only four elements—earth, fire, air, and water. Modern science has shown us that the nonliving world is made of more than one hundred elements and their compounds; these substances are best called "earth" chemicals or *geochemicals*. The nonliving world may be conveniently split into three parts: the solid part or *lithosphere*, the liquid part or *hydrosphere,* and the gaseous part or *atmosphere*.

For millions of years, this was the total world. But about 3000 million years ago a new chemistry developed on the surface of the globe, a chemistry that was to affect the lithosphere, hydrosphere, and atmosphere in a most dramatic way. The new chemicals, called "life" chemicals or *biochemicals*, are all compounds of four of the most abundant elements: carbon, oxygen, hydrogen, and nitrogen. The biochemicals have gradually taken over the earth's surface and have produced the living world, or *biosphere*. That take-over process is what we call *evolution*.

Evolution probably began in shallow pools at the tidal limit of some primeval sea, at the meeting point of the hydrosphere, lithosphere, and atmosphere. Although we know very little about the exact conditions, we do know that the temperature of the sea was below boiling point and that it must have been salty—which means that it was in fact composed of a weak solution of all the soluble elements present in the earth's crust. Any geochemicals present in solution in such shallow pools could be concentrated by evaporation when the tide was out, thus providing ideal conditions for chemical reactions to take place. All that was needed was a big enough source of energy to bring about the formation of new chemicals; in all probability this energy was provided by volcanic activity and by electrical discharges such as lightning. It is believed that it was in this way that the first biochemicals—simple proteins, fats, and sugars—came into existence; compounds that have since been the components of the biosphere.

Life on earth probably began some 3000 million years ago in shallow pools like these formed by tidal waters. Here, where land, air, and salt water mingled, the chemical compounds that became the cornerstones of evolution were formed.

Iceland has been called the land of frost and fire because of its many active volcanoes and geysers. Eruptions like these at Heimaey in the Westmann Islands may well have provided the necessary energy for the formation of the first biochemicals.

In time, the first recognizable organisms evolved. They were nothing more than a unit of biochemicals surrounded by a membrane, and they got their energy by feeding on the dilute soup of biochemicals, which could exist only at sites where the energy from volcanic activity or electrical discharge made synthesis possible. The biosphere could well have fizzled out at this point had it not been for the development of one very special biochemical, which we know by the name of *chlorophyll*.

Chlorophyll is the green pigment that gave the "go ahead" sign to evolution. But why a pigment?

What is so important about color? The problem was that the success of the primitive organisms that absorbed organic chemicals was limited by the amount of chemicals present in the pools in which they lived. Since the volcano/lightning mechanism for producing more organic chemicals must have been both sporadic and inefficient, the organisms' supply of food would soon have been in danger of running out. What was needed was a more widespread source of energy: sunlight.

When sunlight passes through a prism, it is split up into its component colors. Each color is of a certain wavelength, and the amount of energy contained in a given wavelength depends on how long it is; actually, the longer the wavelength the less the energy. The colors we see, that is, the colors of the rainbow—red, orange, yellow, green, blue, indigo, and violet—are visible for the simple reason that these particular wavelengths contain the right amount of energy to cause a photochemical change when they are absorbed by a pigment in the retinas of our eyes. For instance, we see red when a particular object absorbs all the other colors of the rainbow and reflects only the red wavelength.

Early in the process of evolution, an organic molecule was formed that absorbed all the colors in the visible range with the exception of green light, which was reflected. This substance was chlorophyll. Just as the pigment in our eyes, known as "visual purple," brings about the photochemical reaction that allows us to see, so the green pigment brings about the photochemical reaction that underlies the complex chemical process called *photosynthesis* that converts carbon dioxide and water into organic chemicals. The three ingredients for photosynthesis — light energy, water, and carbon dioxide—are in good supply over much of the earth's surface. So once chlorophyll had evolved, there was no holding back evolution.

Of course this did not happen all at once. The evolution of chlorophyll and the other biochemicals that go with it probably took many millions of years. Interestingly enough, it is possible to find the unmistakable bits and pieces that go to make up the chlorophyll molecules in rocks that were laid down over 2000 million years ago, long before we can find any trace of fossils of photosynthetic organisms.

Chlorophyll was therefore the key that opened up the potential of sunlight to the process of evolution. In the early days of the biosphere, the amount of sunlight energy reaching the earth's surface was probably much less than it is today, because there was much more water vapor in the atmosphere. The world was, in fact, wrapped in a thick blanket of cloud. And so there was plenty of water in circulation, as well as a ready supply of carbon dioxide both in the atmosphere and dissolved in the oceans. But there were two important ingredients missing: free oxygen and sufficient nitrogen in a form available to plant life.

Chlorophyll and the process of photosynthesis solved the first of these problems, because photosynthesis makes sugar, and one of the end products of the manufacture of sugar is free oxygen. Take a look at your nearest aquarium when the light is on, and you will see streams of bubbles arising from the surface of the submerged leaves. This is one reason why you need to have plants in

The green pigment chlorophyll is one of the key substances in the process of photosynthesis. No chlorophyll means no photosynthesis, so a quick look at these variegated leaves of a dogwood tree will show you where photosynthesis can take place. The white areas are of little use to the plant.

the tank. They are not there to feed the fish—a daily pinch of fish food does that—but to help supply the water with sufficient oxygen to keep the fish alive. Plants containing chlorophyll have in fact functioned as oxygen factories since the dawn of life, keeping up the supply of this most important gas to both the atmosphere and the surface layers of the hydrosphere.

All plants and animals except a few highly specialized types of bacteria need oxygen for their life processes. For without oxygen, they would be unable to use much of the energy in the sugars made by plants. The process of respiration, which takes place in both plants and animals and releases the stored energy once more so that it can be used for growth and reproduction and other activities, is a process that requires oxygen. Thus in the plant world the two main chemical processes of life—photosynthesis and respiration—balance each other out. The first uses carbon dioxide and water and produces oxygen, whereas the second uses oxygen and produces carbon dioxide and water. They are together a perfect control valve for the biosphere.

At a relatively early stage there was thus plenty of carbon from carbon dioxide, and hydrogen from water, and once there was chlorophyll there was a ready supply of oxygen from the new atmosphere. The one missing element was nitrogen. Somewhat ironically, although 79 per cent of the atmosphere is nitrogen, the majority of living organisms cannot make direct use of the gas because it is a very inert substance. It is a case of "nitrogen, nitrogen everywhere, nor any drop to drink."

Most plants can, however, take nitrogen into their tissues in the form of compounds, especially nitrates. But because nitrates are extremely soluble in water, they are very rare in the surface layers of the earth's crust; they are rapidly leached away by the rain, down into the soil, to springs and thence rivers, and finally out to sea where they form a very dilute solution. What was needed, therefore, was a mechanism that could, so to speak, capture atmospheric nitrogen—the main structural component of amino acids, from which proteins are made—and turn it into nitrate, thus keeping the evolving biosphere well supplied with this essential element.

Among the earliest recognizable plant fossils are simple unicells that are undoubtedly related to present-day bacteria and blue-green algae. So simple is their structure that even with the most modern microscopes it is impossible to recognize any large ordered structures within the cell wall. They are therefore given the special name *procaryotic*, which means "before the cell." Simple in structure these unicells may be, but some of them provided the answer to the nitrogen problem—they possess the unique biological property of fixing atmospheric nitrogen, that is, capturing it and converting it into nitrates. Because they live in the soil they are the source of an invaluable supply of nitrates.

The four main elements of the biochemical cocktail were now readily available to the living system, and the race to tap the potential of the sun's energy was on. The only major obstacle to unrestricted plant growth was the limited supply of certain other geochemicals. This is a limitation that still exists today; to see how it can affect life let us look at the case of *Asterionella*, a plant that

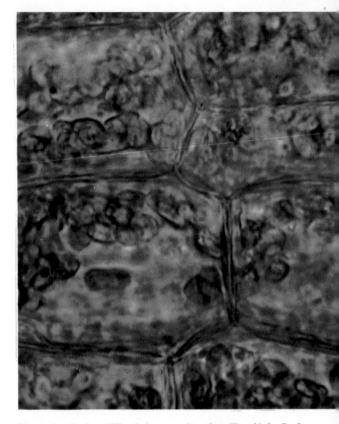

lives in Lake Windermere in the English Lake District. Windermere is a large lake, and *Asterionella* is a very small plant, yet careful study has shown that as the population of *Asterionella* builds up in the lake, it uses up all the available silica from solution. As the supply of silica diminishes, the population stabilizes, and eventually it crashes to a point where only a few *Asterionella* survive. Then, as they reproduce, the cycle starts all over again. Silica is of key importance to this tiny plant because *Asterionella* is a diatom, and diatoms are peculiar among plants in that they have skeletons made of pure silica. Thus, no silica means no skeletons, which means no more diatoms. It is a simple chain reaction: the size of the population is controlled by the resource.

But, whatever the limitations, the biochemicals did begin to produce life some 3000 million years ago in the seas and lakes of this world. It was then and there that evolution started to test out a whole range of simple plant forms. We can collectively call these simple forms the "microscopic algae"—or the "grass" of the waters.

Right: diatoms, such as this one, make up a considerable portion of the plant plankton. Diatoms are characterized by rigid skeletons made of silica, and they take on many shapes and sizes.

12

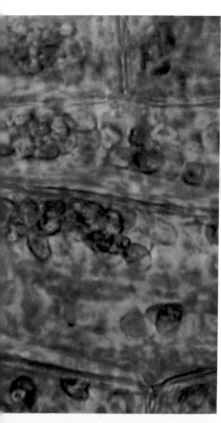

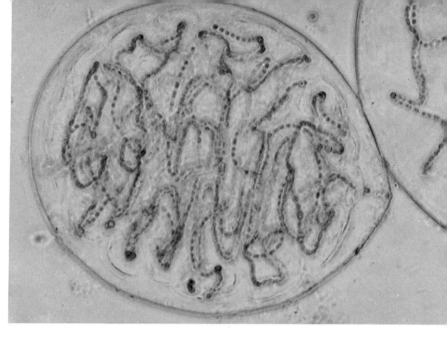

Above: this blue-green alga consists of a mass of many-celled filaments embedded in a blob of jelly. It is one of the few plants that can convert nitrogen gas into plant food.

Left: in most plant cells chlorophyll is found in bodies of various shapes and sizes called chloroplasts. Most common are oval ones such as these (highly magnified).

Below right: the tiny Asterionella *plant is named for the starlike shape of its colonies. The free-floating cells are invisible to the naked eye, but they abound in fresh water.*

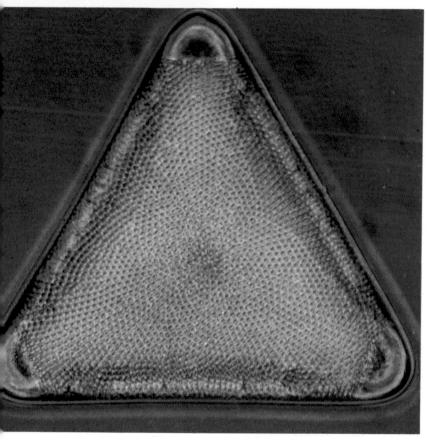

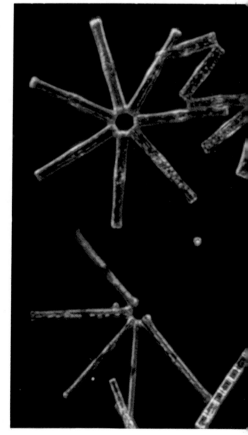

First Plants

Many of the first plants were *planktonic*. That is, they led a buoyant existence, floating in the surface waters of the lakes and oceans. As a result, some of these planktonic plants have long, spine-like appendages that increase their surface area, helping them to float more easily. Success among the photosynthetic free-floaters depends on their ability to keep at the top, because the penalty of falling under the force of gravity is death in the eternal darkness of deep water.

There are problems, however, even for plants that manage to live life at the top, especially if the water is rich in nutrients. The more abundant the nutrients, the thicker will be the plankton soup. And as the soup thickens, the depth to which light can penetrate into the water decreases, so the problem becomes one of keeping right at the very top. Over much of the sea, though, the plankton soup is very thin and light can penetrate to a maximum depth of about 1000 feet. If it were possible to collect all the planktonic plants that live in the lighted zone of a typical area of open sea and to compress them to a solid mass, the resulting plankton biscuit would be only one tenth of an inch thick. Such are the limitations of life in the upper crust of the hydrosphere.

But evolution began in shallow water and it is very likely that because light could reach the bottom, some of the earliest plants became adapted to live there instead of at the surface. So it was that the first real split in the plant kingdom came about —the split between top-level, or planktonic, plants and bottom-level, or *benthic*, plants. The bottom-livers had a whole new set of conditions to meet, not the least of these being the small area that was available to them. Compared with the open oceans, the narrow strip of bare rock situated in shallow water on which plants that need light could live was, and still is, minute. Small as it was, this narrow fringe of illuminated submerged rocks was to be of key importance, for here evolution tried out its "land legs."

Many of the planktonic groups of microscopic algae adjusted to life at the bottom with little

Right: the spherical colony of plant cells known as Volvox *is only about one fiftieth of an inch in diameter. One of the two shown at right is in the process of releasing its daughter colonies.*

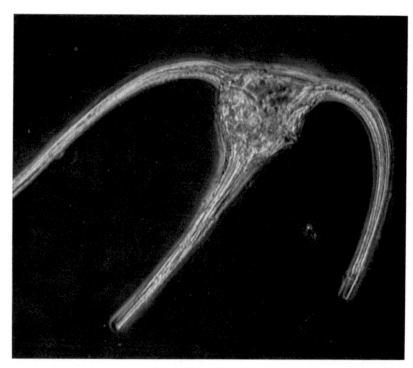

Above: one of the knights of the sea, a dinoflagellate. This small plant rides high in the plankton layer propelled by whiplike filaments that project from its armor plating.

Below: Pediastrum *is a green alga, which forms a colony that looks like a cogwheel. These colonies are found in bodies of fresh water, such as ditches and shallow lakes.*

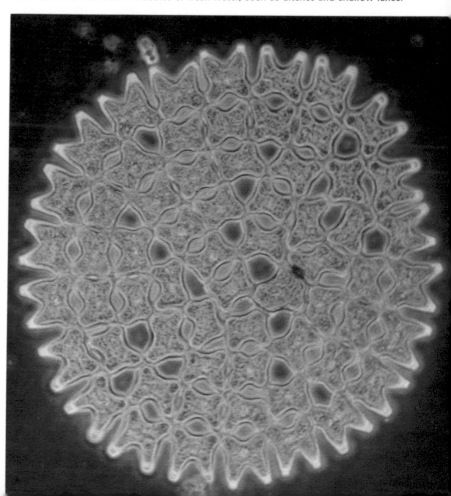

Left: sunlight fades gradually as it penetrates the sea because it is made up of several colors and some of these penetrate farther into the water than others. The diagram at left shows the colors that make up white light (from left to right: red, orange, yellow, green, blue, indigo, and violet) and how readily each color is absorbed both by sea water and the organisms that live in it. The depth to which light penetrates depends on many factors and differs from place to place, day to day, and even hour by hour. The red and orange parts of light are gone at a depth of about 30 feet. Yellow, green, indigo, and violet light penetrate to a depth of about 150 feet and below this only blue continues. This explains why most undersea photographs have a bluish tinge. At about 300 feet even the blue light becomes very faint, giving way to the darkness of the ocean depths.

Because light is very important to the microscopic floating plants that make up the plant plankton (some representative forms are shown highly magnified in the picture above), they occur near the surface of the sea where the light is strongest. The thickness of this "plankton soup" in turn affects the depth to which light penetrates.

Right: Enteromorpha is a green seaweed that inhabits the zone between the tides, being especially abundant in shallow bodies of brackish water.

modification except for the development of a point of attachment, and this was the beginning of a big step forward. One of the key problems of trying to keep up in the plankton is to keep weight down in proportion to surface area; and the easiest way to do this is to remain small. In contrast, once the plant is attached to the bottom, the size/weight ratio is no longer quite so important and it is possible to branch out. The benthic algae did just that. Today, one of the most abundant types of attached algae found in the sea is in fact *colonial* diatoms, the cells of which grow out into long chains instead of breaking free from one another. The brown felt that covers many of the rocks that are exposed when the tide goes out is composed of intricate turfs of filamentous diatoms, some of which grow to more than eight inches in length.

A quick look in a tidal rock pool will show you that three main groups of algae made it in the evolutionary world of the fringe. These were the greens, the browns, and the reds—all of which in time produced the large seaweeds that are such common companions of our seaside holidays. Life is not easy for these plants. We are lucky, for we can choose when to stay on the shore, and generally it is only for a short time each day during a few weeks every year. But the seaweeds have to stay there the year round. They are lashed by storms, exposed to the scorching sun of summer and the biting frosts of winter, and bombarded by surf, rain, and snow. A summer holiday on the beach may be idyllic, but permanent life between the tides is very hard indeed.

The two key adaptations that make the sea-weeds masters between the tides are their *hold-fasts*—the biologist's term for their organs of attachment—and their slimy surfaces. Holdfasts let them hang on tight to rocks, so that they are not washed away, and the sliminess helps to lubricate the fronds, to keep them from getting caught and torn as they are drawn back and forth over the rocks, especially at the turn of the tide. And, too, the sliminess helps to prevent desiccation of the seaweeds at low tide.

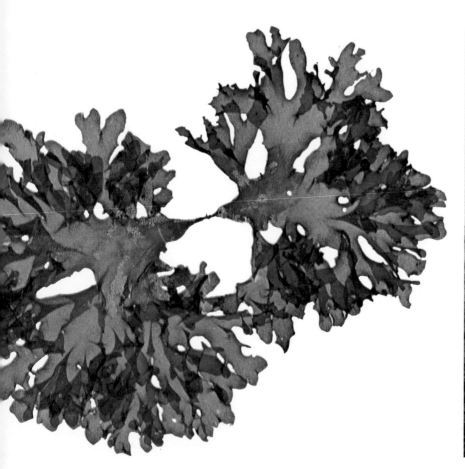

Different seaweeds are adapted for life in a wide range of habitats. Above left: this delicate red seaweed grows in pools on rocky shores. Above right: two common varieties of brown seaweed that are able to survive regular exposure to the air at low tide.

There may be disadvantages in living on the shore, at least when the tide is out, but when it is in, there are few problems. There is an ample supply of water, and because the water is continuously stirred up by wave action, it is well supplied with oxygen. Since the water of the inshore fringe also contains all the minerals that a plant requires for healthy growth, the seaweeds are bathed at least twice a day in a culture medium that they can take in over their whole surface. What is more, the entire surface of these plants contains chlorophyll, and so they do not need a complex internal system of "pipes" to carry water, salts, and sugars from one part of the plant to another. The holdfast not only anchors the plant, but can also take in its own supply of water and minerals and make its own sugars. Large though the seaweeds may be, they are still of relatively simple construction. This is perhaps the main difference between the algae and all other photosynthetic plants: instead of being divided into roots, stems, and leaves, the whole plant body is photosynthetic. It is called a *thallus*, from the Greek word for "twig."

The largest of all seaweeds—and it must be included among the longest of all living organisms—is the giant kelp, which grows in abundance along the Pacific coast of California, where it reaches lengths of over 150 feet. This giant belongs to the worldwide family of kelps, which abound in the cooler waters of both the Northern and Southern hemispheres. The kelps are so large that they appear to need some form of conducting system. The more central cells of the thallus have, in fact, become rather tubelike, though just how efficient they are at transporting substances around the plant is still a matter of conjecture.

There are an enormous number of different seaweeds, but they have not all arisen from the same stock; together with the other types of aquatic plants—the unicellular, filamentous, and colonial forms—they are grouped as "the algae" for con-

venience. Each main group of algae—the greens, the browns, the reds, and the golden browns among others—represents a different evolutionary trend. And together, the members of all these groups have been reaping the potential of the seas for millions of years, and are likely to go on doing so for millions more. But despite their success in shallow waters, it was not the algae that were destined to take the major evolutionary step out onto dry land. This does not mean that there are no algae on land. There are, but all of them are very small, and they either thrive in the wettest places or live an up-and-down existence, growing only just after rain and then drying up to exist in a state of suspended animation until the rain falls again. It was, in fact, other groups of plants, which had been evolving in the shallow waters at the same time as the seaweeds, that had what it took to make it in the dry atmosphere.

In time, the seas, and especially the shallow inshore waters, were teeming with life. Plants and animals of many different sorts were exploiting the varied environments of the ocean. As the potential of the waters became more fully exploited, the problem of competition for resources arose. Demand for nutrients, for space, and especially for light, exceeded the supply, and so the evolving populations of plants were put under stress. The struggle for survival had begun, and nowhere more fiercely than in the cramped world between the tides.

A quick look at any rocky shore in a moderately sheltered position, shielded from direct wave action, will show you that the seaweeds are arranged in zones. Furthermore, each zone is made up of only one type of seaweed. The reason for this seems to be that life between the tides is so harsh that once a seaweed has evolved what it needs to inhabit one zone, it has such an advantage over the others that it can reign supreme. Life in each zone is a real struggle for survival, especially for would-be intruders.

Above the high-tide mark the dry earth was ready for colonization. All the potential—the sunlight, the minerals in the rocks, and the fresh rainwater—was there in abundance, waiting for

Above: the waxy covering that puts the shine on the leaves of the Virginia creeper, helps to prevent the escape of water from the plant. The redwood trees in Sequoia National Park, California (right), are probably the largest of earth's living organisms. Their massive trunks, which may reach a diameter of 25 feet, are protected against water loss, and to some extent against fire, by the very thick bark.

any plant that could overcome the enormous problems of how to cope with life in the thin, dry air. Within the inshore fringe there must have been plants already undergoing key changes in structure and physiology. To colonize the land was but a small step for the plant, yet it was a vast stride for evolution.

Just what were the problems of life on land? First of all, as there is a high percentage of water in every living organism, and because it readily evaporates as vapor into the air, one immediate problem was to control loss of this precious liquid. What was needed was a watertight skin—a layer of wax. Glance at any nearby leaf, and you will see the shiny surface layer that was evolution's answer. The shine on the leaf surface is the *cuticle*, and it is made of a waterproof, waxlike substance known as *cutin*. But the cuticle itself would present a problem if it were thoroughly impermeable, because a plant's green photosynthetic cells need a regular supply of carbon dioxide, and that has to get in somehow. The answer evolved by land plants is the pores that are present all over the surfaces of their stems and leaves. Each pore has an aperture that can be opened and closed to control water loss and carbon dioxide intake. These pores are called *stomata*, and they are the stamp of all true land plants. They are harder to see with the naked eye than the overall cuticle, but you can just spot them with a modest hand lens, and it is surprising how many there are. The undersurface of an average leaf has many thousands per square inch.

Problem number two arose from the fact that whereas a water plant is bathed in its "culture medium," which it can take in over its whole surface, a land plant can obtain the water and minerals it needs only from the soil. It therefore requires a means of penetrating down into the damp substratum. The holdfast of seaweeds is a good anchor, but it lacks the property of penetration. To live on land, plants need roots. But although you might think that the development of a separate shoot system and a subterranean root system should have given the land plants instant terrestrial bliss, this was still not enough. Every part of a plant performs a special, separate function, and so there was now the need for a system of "pipes" to provide rapid transport from one functional part of the plant to the others. In short, the land plant required a vascular system.

The third problem was one of support. An aquatic plant is supported by the water in which

it grows, but thin air cannot serve such a supporting role. To keep the plant body from collapsing, strength had to be built into it. And so, in time, woody tissues evolved—tissues that were to support even the might redwood trees of the Californian coastal range.

The greatest problem of all, that of how to breed out of water, was not satisfactorily solved until quite late in the long course of evolutionary history. The reproductive stages in the life cycles of most algae are free-swimming. And the most crucial phase in the whole life cycle is the delicate task of getting the free-swimming male reproductive cell to the egg so that fertilization can take place. Without the process of sexual reproduction, no exchange of genetic materials could take place, and evolution would probably have ground to a halt. Here was the critical problem that could have thwarted the evolution of successful land plants.

The key to the solution of the problem was the splitting of the life cycle into two very different phases or generations: the *gametophyte* generation bearing the organs of reproduction and needing free water for transport of the male cells to the female egg cells, and the *sporophyte* generation producing spores resistant to drought. As we shall see in the next chapter, it was the sporophyte generation that developed into the familiar land plant with its complete freedom from any need for free water for reproduction.

It was the French botanist Lignier who put forward the idea that the direct ancestors of the first land plants were dichotomously branching (repeatedly forked) algae, which probably looked not unlike the bladder-wrack seaweeds. Lignier believed that this ancestral plant lived along the shoreline where intermittent drying of the habitat promoted the evolution of the right adaptations.

Only 14 years after Lignier had put forward his theory some very exciting fossil evidence came to light. The fossils were found in blocks of a flintlike rock called chert that had been used in the construction of a wall. The origin of the rock was traced to a quarry near the small village of Rhynie in Scotland and the fossil plant were therefore called *Rhynia*. The chert had been laid down about 300 million years ago, probably in fresh water, so *Rhynia* may have been semiaquatic. Nevertheless

Before the plant kingdom spread from sea to land, much of the earth's surface was completely devoid of life and may have resembled this barren Arizona landscape.

it had much of what it takes to make a land plant and predates all other fossil evidence of a land flora. The fossils were in fact so well preserved that the conducting tissues, the *epidermis* (outermost cell layer), the stomata, and the resistant spores were easily recognized. In general shape *Rhynia* looks not unlike a stiff, thin bladderwrack seaweed, but in all other respects it most closely resembles the Pteridophytes—a group of plants that includes the modern ferns.

Here, then, is one of the major gaps in our knowledge of the evolution of the plant kingdom. The links between Lignier's seaweed and *Rhynia* are completely missing. All we can conclude is that *Rhynia* is a good prototype land plant that has the majority of the modifications possessed by contemporary land plants. We must continue to hope that one day the missing links will be found. It is fascinating that, even today, two plants that look very like—and are indeed related to—*Rhynia* can still be found growing in abundance in the tropics. These are *Psilotum triquetrum* and *Psilotum nudum*. They may well prove to be the direct descendants of *Rhynia*.

There is no evidence, fossil or otherwise, to suggest that this type of plant has ever been a success on a world scale, for the fossils are not common. However, it depends on one's definition of success. The basic format of these prototype land plants has certainly survived almost unchanged for over 300 million years of evolutionary time, which is not a bad record of fitness.

Although we may never know exactly what type of plant took the first "small stride," we do know that once all the problems had been solved the bare landscapes of the world were quickly clothed with a living mantle of green, and the potential of the whole earth was opened up to evolution. Even now, when the great Arctic ice sheet relents a little, melting to expose new bare landscapes to colonization by plants moving up from the warmer south, the first plants to grow are lichens, mosses, and liverworts. And as we shall see, these plants occupy a position midway between water and land plants. Thus it is that even in our own time, the age-old story of the colonization of the land is acted out in the same way again and again.

Left: Psilotum nudum *is rightly called a "living fossil." It consists of angled green stems that have no true leaves, but only minute scale leaves at the base of each yellow spore case.*

Opposite page: the illustration shows the solutions developed by land plants in response to the problems faced by the first plants that ventured onto land. Most immediate of these problems was the control of water loss from the plants.

Problems Related to Life on Land

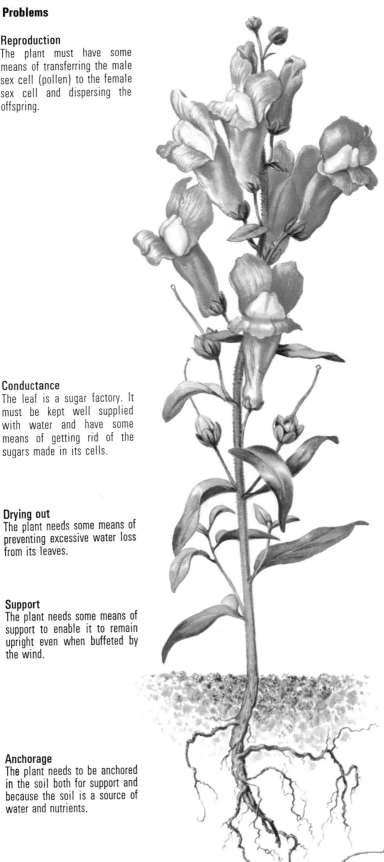

Problems

Reproduction
The plant must have some means of transferring the male sex cell (pollen) to the female sex cell and dispersing the offspring.

Conductance
The leaf is a sugar factory. It must be kept well supplied with water and have some means of getting rid of the sugars made in its cells.

Drying out
The plant needs some means of preventing excessive water loss from its leaves.

Support
The plant needs some means of support to enable it to remain upright even when buffeted by the wind.

Anchorage
The plant needs to be anchored in the soil both for support and because the soil is a source of water and nutrients.

Solutions

Development of brightly colored, scented flowers with nectar, which attract insects that carry pollen from plant to plant. Fusion of sex cells results in the production of seeds that are resistant to water loss. Seeds are dispersed by animals, or wind, or sometimes by water.

Development of a conducting system to carry water and salts up to the shoot. The conducting system also transports sugars from the leaves to the rest of the plant, and it conducts hormones for the control of plant growth.

The leaves develop waxy surfaces with pores that allow gas exchange and control water loss from the plant.

The conducting system gives strength, but also allows some flexibility so that the stem can bend with the wind.

Roots fulfill both these needs. The conducting system in roots is concentrated as a central rod so that the root has extra strength for pushing into the soil.

25

Living
Experiments

If you were to ask me to name the commonest alga that lives out of the water, exposed to dry air, I should find it impossible to give you a straight answer. The blue-green alga *Nostoc* must rank among the top ten—although most people will never have seen it, for the simple reason that it is small and of an insignificant dirt color, which makes it difficult to distinguish from the soil on which it grows. But it is present on most soils and, because it is a nitrogen fixer, is of great importance to the world at large. The most conspicuous land alga is probably the green one that covers tree trunks and fence posts; this single-celled plant, called *Protococcus*, lives a very up-and-down sort of existence. Gray-green when dry, with its life processes almost at a standstill, it "waits" for rain; and when rain comes, all its systems are

"go," and active growth starts again. If the dry period lasts too long, that particular generation of cells will die, but the life of the plant continues within special thick-walled cells called *spores*. These spores are resistant to desiccation and may lie dormant for very long periods.

One of the greatest truisms of biology is that with the evolution of life came death and dead organic matter. Dead organic matter is potential for any form of life that can make use of it. Bacteria and fungi are colorless plants that evolved in response to such potential. Although many of them are in fact deadly parasites and get their nutrients from living plants and animals, many more are *saprophytes* (meaning literally "eaters of rotten matter"), and they make use of the energy in dead organic material, cleaning up the environment in the process. If these decomposers had not evolved, evolution would soon have come to a halt, choked by its own debris.

The best way to see fungi is to take to the countryside and look for those weird but familiar toadstools. Although they get their name from the German *tod stuhl*, meaning "stool of death," they include among their numbers not only some of the deadliest but also some of the most delicious plants to be found on earth. The mushroom or toadstool that we eat, or leave well alone, is in fact only a minute part of the plant; the main part grows underground and gets its nutrients from dead organic matter in the soil. Perhaps the best way to understand fungi is to grow them on a piece of moist bread under a glass cover. It requires a little more patience these days than it used to, because the recipe of modern mass-produced loaves includes chemicals that inhibit the growth of fungi. In time, however, a thick network of white threads begins to appear on the surface of the bread and to permeate down into it. These threads are *hyphae*, which is the name used to describe the chains of cells that make up the body of the fungus.

Very soon the surface becomes discolored by black, yellow, red, or blue dots. These are the reproductive bodies, which produce long-lived, resistant spores. If you allow the bread to dry up, so will the mold; in drying up, however, it releases millions of spores that can travel considerable

Any piece of terrain left untended will become covered with vegetation—even if the terrain is only a bowl of stale food. Above: molds grow in profusion on rice pudding. Right: a conspicuous green alga colors the roots and branches of a yew tree and caps the tops of nearby rocks.

distances—which is how the fungus got to your bread in the first place. Fungal spores have been recovered from the upper atmosphere in a viable state; so, when you have finished your demonstration, make sure you destroy the fungal garden!

Wherever there is organic matter there is an array of saprophytic fungi living on this organic matter. Through their life processes these fungi help to keep the environment clean and healthy, and they recycle minerals back into the soil to nurture other plants. Whether large or minute, exotic or ugly, each fungus performs its own task in the ordered web of life. And it is a fact that order may be found where at first sight there appears to be chaos. Careful study of the development of a fungal garden will show you that the succession of different fungi, often distinguishable by their different-colored spores, occurs in a regular pattern. The steps in this regular succession of fungal types are brought about by competition for resources and for space.

Perhaps it was competition that helped evolution to take its first step on to dry land, but it was to be cooperation that would eventually open up the full potential of the dry earth. Competition can be defined as mutual interference among plants, cooperation as mutual help. And without doubt, the best example of mutual help is the lichens, which are abundant members of many types of vegetation. Like the blue-green algae, many of them are small and insignificant, especially when growing among other plants. In certain types of vegetation, such as tundra and humid forest, they can, however, play a prominent role, giving their own special look to the vegetation.

Lichens come in all shapes and sizes. The smallest are the minute pyrenocarps, which appear as black pinpricks scattered over the surface of rocks, especially limestone. The black pinpricks are in fact the minute lichen fruits, which occupy tiny cavities in the rock surface. Each fruit contains masses of lightweight spores that are dispersed by the wind.

At the other end of the size scale are the large, flamboyant fruticose lichens that hang from trees in the forest and can reach lengths of more than 20 inches. My favorites, though, are the crustose lichens, which weave intricate and often brightly

The bright orange growth on the rock in the foreground is a living skin of lichen. The dual tissue of alga and fungus of which lichens are composed eats down into the rock.

colored patterns on the surface of rocks and are constant companions of any mountain walk.

Whatever their size, lichens all have the same basic components. They are composed of a mixture of mutually dependent algal cells and fungal hyphae that are interwoven and interact with each other, producing these weird dual plants, which inhabit some of the harshest environments on earth. Both the fungi and the algae that go to make up lichen bodies are delicate cells that on their own would rapidly dry on exposure to air. Together they interact, the fungal partner producing a tough structure that is resistant to water loss, within which the algal cells are protected. The fungus forms the "house," and the algal cells provide the fungus with at least some of the organic nutrient it needs.

We have not yet begun to understand how two

Two lichens are shown in the picture above: a whitish-gray one with stalked cuplike structures, and a dull yellowish-green one with clusters of bright red spore-producing structures.

The Development of the Sporophyte as the True Land Plant

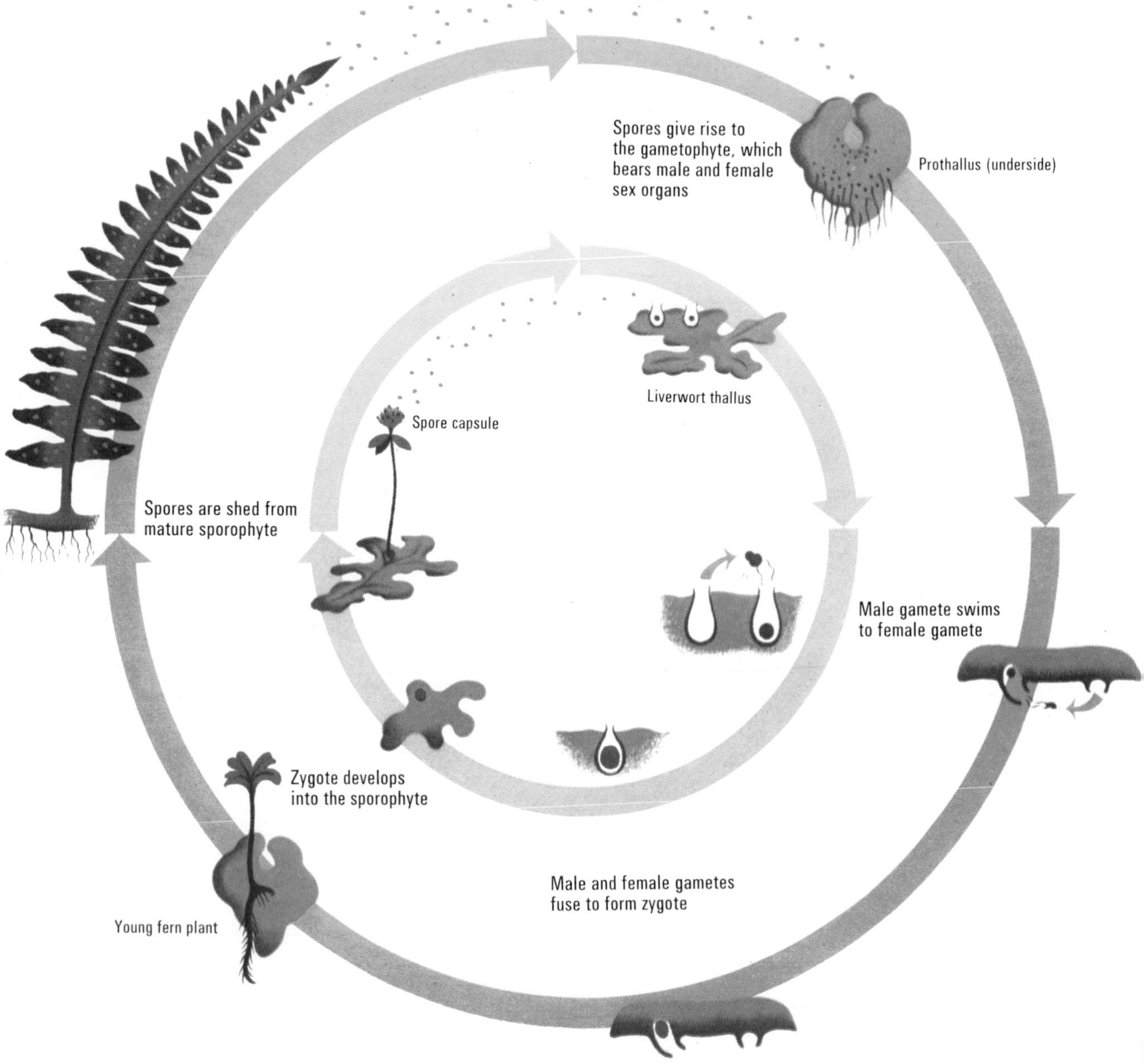

Spores give rise to
the gametophyte, which
bears male and female
sex organs

Prothallus (underside)

Liverwort thallus

Spore capsule

Spores are shed from
mature sporophyte

Male gamete swims
to female gamete

Zygote develops
into the sporophyte

Male and female gametes
fuse to form zygote

Young fern plant

completely unrelated plants, each of which has its own form, can interact to produce a third, completely different, form. But it was way back in 1886 that the botanist Gaston Bonnier first claimed to have made a lichen by mixing a certain kind of fungus with a certain kind of alga. Although Bonnier's claim is suspect because it has never been repeated under strictly controlled conditions, it did lay the foundations for a long and detailed study of this extreme form of cooperation,

In the struggle to overcome one of the major problems of dry-land living—that is, the problem of breeding out of water—the life cycles of land plants split into two different phases or generations: the generation producing the sex cells or gametes (the gametophyte), and the asexual, or spore-bearing, generation (the sporophyte). The life cycles of a liverwort (inner circle) and a fern (outer circle) shown above illustrate this splitting into two alternating phases. In both plants the gametophyte (colored green) needs water for the male sex cells to swim to the egg cells, while the sporophyte (colored brown) is not dependent on water in this way. The fern life cycle indicates the evolutionary trend toward the sporophyte as the dominant phase with the gametophyte reduced to a minute prothallus.

known as *symbiosis*. The fact that it happens is of great importance to the two component plants. Alone, neither plant could colonize the drier habitats, but together they form a simple plant cocktail capable of growing on exposed, inhospitable rock surfaces. Lichen growth is, of course, still controlled by water supply; but not to the same extent as the component algae and fungi when growing on their own as independent plants.

Not only do lichens open up the potential of the rock surface to their component parts, but, as we shall see in later chapters, they also help to prepare it for colonization by other plants. By their death and decay the lichens add organic matter, which holds water and minerals, in a form in which the minerals are available for use by other plants and animals.

Wherever you find an abundance of lichens, you will usually find mosses and liverworts growing nearby. These are two groups of plants that have, for a long time, been erroneously linked together under one name, the Bryophytes. Their only real claim to connection, however, is that both have overcome some of the problems of living in dry air by adopting the "schizophrenic" way of life — splitting their lives into two very different phases. The most conspicuous phase of the life cycle is the gametophyte, which bears the sex organs. This lives the on/off type of life tied to the wet substratum that is ideal for those mobile male reproductive cells, the swimming *antherozoids*. The gametophyte alternates with the spore-bearing sporophyte generation. Both phases usually have chlorophyll and are photosynthetic, but only the sporophyte, which grows up into the dry air, has stomatal pores, to control water loss from its tissues.

Liverworts come in two basic sorts: the simple ones, which look not unlike the lobes of a liver and gave the group its common name, and the more delicate leafy ones. It is easy to tell that a leafy liverwort is not a moss, because its leaves are arranged in two ranks, one on either side of the stem. In contrast, moss leaves are arranged spirally around the stem, making it very difficult to see the stem among the overlapping leaf bases. The liverwort leaves are also often divided and lobed, so that they can be very complex structures. The leaf of the liverwort *Frullania*, which forms dark, red-brown patches on wood and rocks, especially in more humid coastal regions, is among the most bizarre. Helmet-shaped lobes on the underside of the leaf act as minute reservoirs, helping to conserve water for the plant, tiding it over short, dry periods at least. However, another plant has taken advantage of the situation and lives in the reservoir. It is a blue-green alga that is able to fix nitrogen, thus aiding the liverwort. This is, in fact, another case of mutual help.

Because the mosses and liverworts are similar in so many ways, it is perhaps easy to see why early botanists placed them in the same group of the plant kingdom. More detailed study has shown that this is incorrect; both are living experiments that have become fitted by their evolution to exploit the betwixt-and-between world; they have

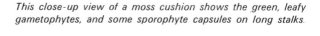

Pellia, a liverwort. The green-lobed gametophyte produces the sex cells that give rise to the black sporophyte capsules.

This close-up view of a moss cushion shows the green, leafy gametophytes, and some sporophyte capsules on long stalks.

got something of what it takes, but cannot be
regarded as true land plants. The fascinating
thing is that although we can learn a lot from the
mosses and liverworts about how the plant kingdom overcame some of the problems of dry-air
living, there is no evidence to suggest that the
evolution of the land plants took place via the
Bryophytes. Whatever the missing link, evolution
was working on the spore-bearing generation,
for only this possessed the stomata so essential
to the plants' success on dry land. From this
point on in evolution the gametophyte generation
slowly lapsed from being the dominant phase
until, as we shall see, nothing but the actual
process of sexual reproduction was left. In this
generation game the gametophyte lost all along
the line, the sporophyte becoming dominant in
the drive to exploit the potential of the dry land.

At the same time as the lichens and Bryophytes
were testing out their land legs, another group of
plants was evolving—a group that may collectively be called the Pteridophytes. This group
includes the modern ferns, clubmosses, horsetails
and *Psilotum*, all of which are true denizens of the
dry land, although, as we shall see, they are still
limited by that water-demanding gametophyte.

One of the commonest ferns is bracken, which
is a serious agricultural pest, especially in rough
grazing land. The great fronds, which can top
seven feet in height, are compound leaves—
complex organs covered with a thick cuticle and
well furnished with stomata. The fronds produce
not only the sugar needed for the growth of this
large plant, but also many thousands of spores,
which develop in rows of *sori* on the back of each
leaf. This sporophyte generation possesses all the
attributes of a land plant: root, stem, leaves, and
an internal conducting system. Out of the millions
of spores produced by a mature bracken plant,
those that fall on good ground germinate to produce, not new bracken plants, but small green
scales called *prothalli*. These are so small that it
requires careful searching of the damp, shaded
earth around a bracken colony to locate them.
However, small though they may be, this is the
other half of the life cycle—all that is left of the
gametophyte generation. The prothallus bears the
organs of sexual reproduction, and the antherozoids, once released, can swim through the film of

*A stand of bracken growing in the autumnal shade of pine
trees. Soon each frond will release thousands of tiny spores.*

The Interrelationships of the Plant Kingdom

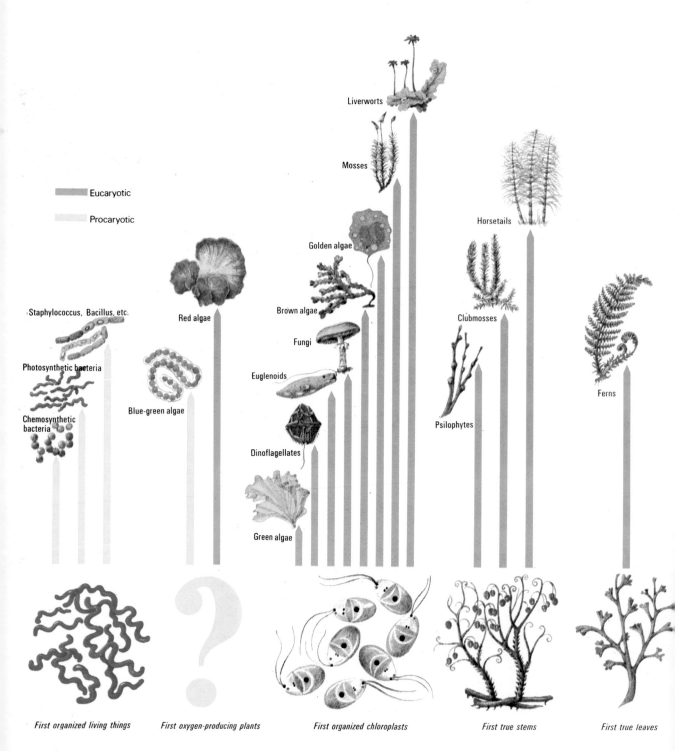

Eucaryotic

Procaryotic

Staphylococcus, Bacillus, etc.

Photosynthetic bacteria

Chemosynthetic bacteria

Red algae

Blue-green algae

Golden algae

Brown algae

Fungi

Euglenoids

Dinoflagellates

Green algae

Liverworts

Mosses

Horsetails

Clubmosses

Psilophytes

Ferns

First organized living things   *First oxygen-producing plants*   *First organized chloroplasts*   *First true stems*   *First true leaves*

water that covers everything down at "prothallus level." It is very difficult for human beings to imagine what it must be like to be so small that a drop of water becomes an Olympic-sized pool.

This unequal development of the two generations was the great breakthrough that made it possible for large plants to evolve to exploit the fringe world between water and dry land; and it was plants of this type that made up the vegetation of the great swamp forests of the Carboniferous period. Coal is the fossilized excess of millions of years of photosynthesis by a whole host of these plants—the giant Pteridophytes, all now extinct.

The format of success, which enabled the first large plants to lift their leaves up toward the sunlight, was a root-and-shoot system, interconnected by a set of pipes that both gave the plant the necessary strength and allowed the transport of water and nutrients. A pipe capable of conducting water must be hollow, waterproof, and rot-proof, and thus the plant cells that take over the functions of water conduction must die in the execution of their duty. This is because living plant cells are no more than a dilute solution of living chemicals in water, so there would be an enormous problem of controlling the continued dilution of the living chemicals composing the cell. The system of dead cells that was evolution's answer to the problem is called *xylem*, or wood. As the particular cells that are destined (by their position in the developing plant body) to be part of the water-conducting pipe elongate, their walls become impregnated with a chemical called *lignin*, which is both a waterproofing agent and an inhibitor of bacterial growth. Once the cell has completed its growth, it dies, partly as a result of the deposition of lignin and, in death, fulfills its two important roles— water conduction and support.

Increase in height must go hand in hand with increase in girth—or, at least, considering the use of steel girders to support tall modern buildings, there must be an increase in support in relation to total weight. In plants this is brought about by the formation of more dead wood each year. It was this living death that allowed evolution to lift the chemistry of life up above the earth toward the potential of the sun. In every tree, the bulk of the structure consists of dead cells that have at some stage carried water up to the leaves above. These dead cells are closely associated with a complex of living cells, each of which performs some special function in the life of the tree. For instance, the cells of the *phloem*, which lies

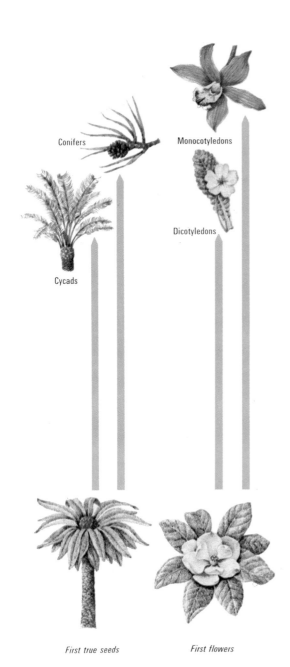

Conifers

Monocotyledons

Cycads

Dicotyledons

First true seeds   *First flowers*

just beneath the bark, surround the wood and carry the sugars around the plant; some living cells among the dead cells of the wood store food, and there are others in which waste materials are "filed away"; some cells divide actively each year and form new layers of wood, whereas others divide to form new layers of cork to protect the expanding trunk. All these cells together form the trunk of a living tree, which in more ways than one is the height of plant evolution.

The main constructional biochemicals of plants are cellulose and lignin, and the backbone of both of these is carbon, which is derived from the carbon dioxide of the atmosphere. Lignin is very special in that it has bacteriostatic properties. Among the crowded fossils of the giantic extinct ferns are the faint impressions of what were undoubtedly fungi. In the swamp conditions of the Carboniferous period, however, there was little likelihood of these decomposers breaking the plant material down, because the swamp waters were devoid of oxygen. It is hard to say whether the bacteria-inhibiting action of the lignin played any significant role in preventing the giant plants from being totally decomposed, but the result—coal—is unquestionable and there is still enough of that in the earth to prove that the remains of the coal-forest plants underwent only partial decay.

In this way, an enormous amount of carbon was locked up, and it has been estimated that the present day atmosphere contains only one fifth of the amount of carbon that was locked into the world's coal deposits. It is interesting to speculate whether this put the current account of carbon dioxide in the red, or whether the mass loss from the atmosphere was compensated for by the release of the gas from the sea. Recent measurements indicate that the level of atmospheric carbon dioxide is at present on the increase, and at least some of this is coming from the combustion of fossil fuels. In other words, in our furnaces and motor cars, we are in fact completing the job that was started by decomposers such as the fungi, over 750 million years ago.

If the present day situation is in reality as I have suggested, it could be argued that the Carboniferous period, which was the great period of photosynthetic affluence, was brought to an end, at least in part, by lack of carbon dioxide. It could also be argued that as the level of carbon dioxide rises, we are heading for another period of botanical affluence.

Whatever the reason, the Carboniferous period did come to an end and the giant "fern" forests disappeared from the swamps of the earth. Meanwhile, in those same swamps, two other groups of plants—the cycads and the conifers—had been evolving toward the final break with the water. Fantastic as it may seem, we do know something even about the prothalli of the giant Pteridophytes of the past. Fossil evidence has shown us that they were much like their modern counterparts: small, delicate structures bearing the reproductive organs on their lower surfaces. It seems safe to conclude that they lived a similar life, confined to the layer of humid air over the damp swamp soil. The Pteridophytes could not, therefore, evolve any further as long as they were stuck with their delicate prothalli. The prothallus, or rather the free-living prothallus, had to go; and so evolution overcame the problem by enclosing the prothallus in the sporophyte plant. The best way to understand how it happened is to look at the cycads.

Cycads look not unlike ferns stuck on the top of tree trunks. Close examination, however, immediately shows you that they are not. The young leaves do not emerge rolled up like a bishop's crosier, and when mature, their backs are not covered with spore cases. Cycad spores are, in fact, borne in gigantic cones, which may be almost three feet long, and are composed of a central axis bearing spirally arranged scales. The cones are of two sorts, male and female. The male cones produce small spores, which can be readily transported by the wind. The larger female cones contain and partly enclose large, often colored ovules. The male spores, which are best called pollen grains, are blown by the wind, and some get caught in among the scales of the female cone. At the top of each ovule is a small hole, into which the pollen grains are drawn. Each pollen grain germinates to produce a pollen tube that grows down toward the egg cells, but in each ovule only one pollen tube can win the race to each of the egg cells.

The developing pollen tube contains some of the weirdest antherozoids in the business. They look not unlike wooden whipping-tops, even down to the spiral groove on which you wind the string!

Cycads are primitive seed plants that bear separate male and female cones. This loosely-packed female cone (above) is composed of pinkish-colored ovules borne on leaflike structures.

The string itself, though, is a row of long flagella, which can spin the mini-top to its destination once it is released into water. The ovule contains a small pad of tissue—all that is left of the female prothallus—enclosing two egg cells, which lie at the bottom of a private indoor pool (the pollen chamber), the water in which is produced by breakdown of part of the tissue. It is through this that the antherozoids swim to complete the process of fertilization. But that is not the end of the story. The fertile egg develops within the protection of the ovule and the cone, producing the embryo of a new plant. The whole thing—embryo, plus food store, plus protective coat—is, in fact, a seed (the perfect package-deal reproduction kit for life on land). The cycads are the most primitive living seed plants and, hence, the most primitive of the real land plants. They have roots, stems,

leaves, a vascular system, and a seed. With none of the problems of a delicate, free-living prothallus, they are no longer tied in any way to a damp habitat.

Once the first seed plant had evolved, there was nothing to stop the invasion of the dry earth by life. And this seems to be as good a point as any to review the whole of the plant kingdom.

It is easy to see both the structural and functional relationships between the great cones of the cycads and the smaller, much more familiar cones of the conifers, such as pine, spruce, larch, and juniper. Although most people have seen pine cones, very few bother to look for the seeds. The best way to see them is to collect a fresh cone and stand it in a dry place. Gradually, the cone scales open in response to the drying to reveal the seeds, which are often winged so that they will be dispersed readily by wind. In the conifers, the breakaway from dependence on water is even more complete than in the cycads; the pollen tube grows down to deposit the male cells very close to the receptive nucleus of the egg; there is no need for swimming antherozoids nor for the private pool.

It is a great shame that every family does not possess a microscope. An evening looking down even a moderately powerful microscope can be an experience never to be forgotten. The Victorians were great microscope gazers, and most well-to-do households would draw up their chairs while father demonstrated the wonders of everyday things, with the help of a little bit of magnification. One of the highlights of an evening's viewing would be to take pollen, place it on a slide in a drop of dilute sugar solution, and watch what happened. I have sat for hours watching pollen tubes, and the whole process will never cease to amaze me. Very soon, the pollen grain splits, and a tiny, transparent tube appears and begins to grow rapidly, snaking its way across the field of view. A few hundred grains on the slide—and the result looks just like the writhing head of the mythical Medusa. It is enough to give some people the "itches," especially if they have a tendency to suffer from hay fever.

Hay fever is an allergic reaction to tiny particles that are breathed into the respiratory tract of the sufferer, and pollen grains are among the worst offenders. If it hadn't been for the evolution of those seed plants whose pollen is dispersed by wind, just think how much more pleasant summer might be, at least for some people. The millions of pollen grains that make up a large part of the daily spore count, which is today as much a feature of our summer weather forecasts as are the isobars, are derived from the whole cross section of types of seed plants: cycads, conifers, and flowering plants. In any one area, the most abundant grains will depend on which species of plant is growing in the near vicinity, but, of course, some grains do get carried for enormous distances.

The end of the current line of evolution of the land plants is the Angiosperms (whose name, derived from a Greek word meaning "vessel" or "container," indicates that the whole reproductive process is contained within the ovary and seed coats). The containers that hide the process from our view are, of course, flowers—the organs of reproduction that are among the most beautiful objects on earth.

Close inspection of any of the larger flowers will show how they resemble, at least in structure, the cones of the cycads. The various parts of a flower—the sepals that protect the delicate flower-bud; the

Like all flowers, this scarlet pimpernel is a complex organ of reproduction. Its red petals, which are closed in rainy weather, open to attract insects that pollinate the flower.

showy petals, whose function is to attract insects, not man; the anthers, with their pollen sacs; and the carpels that enclose the ovules—are all either reduced leaves or buds that have been modified for the process of reproduction. One of the first people to realize this was the 18th-century German poet and dedicated botanist, Wolfgang von Goethe.

It is worth keeping your eyes open in the countryside and looking for flowers that have in part reverted to the original form. Instead of petals, the flower is made up of whorls of tiny leaves, proving that Goethe's surmise is correct. Such leafy flowers are much harder to find than four-leaf clovers, but much more exciting; the scarlet pimpernel is a good plant on which to start searching.

Two families of the flowering plants have

perhaps gone the furthest into the insect pollination business. One is the orchid family, which everyone must know about; the other is the Asclepiadaceae. When most of us think of orchids, it is of the large, showy, hybrid ones that have been produced by the horticulturalist for hothouse use and are derived from tropical ancestors. But just as beautiful are the smaller wild orchids of the temperate region, and it is these orchids that really show the extent to which flowers may be dependent on insects for their pollination, and hence for their very existence. Many of the orchid flowers bear a strong resemblance to the insects that pollinate them—the fly and the bee orchids being perhaps most famous. So nearly perfect is the resemblance of the orchids to insects, that insects do, in fact, try to mate with them. The result is that the orchid pollen grains, which are borne in two special sacs (the *pollinia*), come away attached to the insect, which carries them intact to the next flower.

The relationship is even more extreme in the Asclepiadaceae family. Not only does the insect have to carry the pollen mass from one flower to

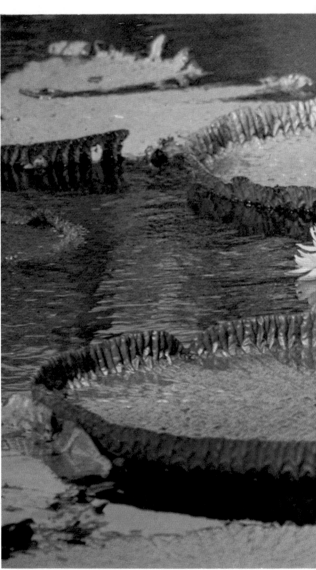

Although the flowers on these pages vary in appearance, they have evolved in response to the same problem—that of attracting insects to the plant so that pollination and thus seed production can occur. The waxy petals of the magnolia (left) hide numerous ovaries, each containing a single seed.

another but each pollen mass has a specific shape that corresponds to a hole in the stigma of the recipient plant. There is no evidence to suggest that the insect actually has to fit the key into the keyhole, but this lock-and-key mechanism certainly helps to ensure that only the right pollen gets to the right ovule.

At the other end of the line are the flowers of the grasses, most of which hardly deserve to be called flowers at all. In the grasses there is no need for great, showy petals, because pollination is effected by the wind. One reason for the success of the great family of grasses is the fact that wind pollination is a very efficient process. But though highly efficient, it is not very selective, and thus pollen grains fall quite haphazardly onto different flowers. There must, therefore, be some way of making sure that fertilization cannot take place between the wrong plants. If it did, there would be a continuous remixing of all the genetic characteristics, and evolution would not have got very far. As it is, evolution of new types of grass must have gone hand in hand with the development of barriers that prevent interbreeding.

Above: the giant Amazonian water lily has enormous floating leaves and large white flowers. The showy blooms are essential to seed formation because they serve to attract the insects that pollinate them. Right: a fly can easily be tricked by the flower of the fly orchid, which is such an excellent imitation of the insect itself that flies attempt to mate with it, and this results in the flower's pollination.

The sensitive Mimosa pudica *(from the Latin* pudere *meaning "to be modest")* is a tropical weed that reacts to being touched by folding up its green leaves and playing dead. The picture above shows an undisturbed plant with its leaves erect; above right, the same plant in its state of modest collapse.

Right: pampas grass takes its name from the plains of South America. It is a densely tufted perennial grass, which produces glossy white plumelike flower heads on stalks that may reach a height of 10 feet. Unlike most grasses it has rough-edged leaves that make it very unpalatable to grazing animals.

The cereals are a classic example of this. They are all self-pollinating; and because each flower pollinates itself there can be no mistake. From the evolutionary point of view, however, this can be just as bad, because self-pollination makes the mixing in of new characteristics impossible. Modern plant breeders have surmounted this obstacle by artificially crossing different sorts of cereal. For the most part, the results have been nothing out of the ordinary, but occasionally a plant breeder has hit the jackpot. The best example was the production of the super cereals. These are varieties of cereal that have such large flowering and fruiting heads that the amount of grain they produce is nearly double that produced by standard cereal plants. It is these new high-yield cereals that form the stock-in-trade of the "green revolution"—the contemporary agricultural revolution that could go a long way toward solving the world's food problem.

Unfortunately the flowering plants, growing as the majority of them do on dry land, where their remains are exposed to the oxygen-rich air and thus decay rapidly, have not left a detailed record of their evolution as did the great Pteridophytes. It is interesting, therefore, to speculate on how many weird forms of plants have lived, had their day, and passed on, leaving no trace. And why have some equally strange forms survived? What was it, for instance, that led to the evolution of such sensitive plants as *Mimosa pudica*, which folds up when touched? This is a common roadside weed in parts of the tropics; and it is a peculiar feeling to drive down a road that is edged with green, only to turn and see that the edges that are covered with the sensitive mimosa appear to be brown and dead because the plant has folded up its green leaves. There seems little doubt that this reaction of the sensitive plants could have some survival value in that it discourages grazing animals, but I always wonder exactly what the plant's origin was, and whether there have been other, larger, sensitive plants in the past.

Most peculiar of all flowering plants is the telegraph plant, which must be the most mobile of all. When the sun is shining and the temperature is above 70° F, the leaflets and leaves of this plant are all of a twitch. To watch a stand of *Desmodium gyrans* is very like watching a crowd gyrating at some discotheque. It is difficult to think of what function this vegetable twist could possibly have. Perhaps it serves no useful purpose at all!

Plant Communities

It was through the process of evolution that plants gradually developed the adaptations that allowed them to exploit the varied environments of this earth. The successful mechanism was the natural selection of particular traits that developed in certain members of the population—traits that gave them some measure of advantage in a particular environmental situation. In the sea, for example, if a member of the plankton developed a more efficient mechanism for floating, it would then ride high among the free-floaters, getting the full benefit of the sunlight. If it could pass that particular trait on to its offspring, they would start life with the same "upper crust" benefits, and would themselves have a better chance of reproducing and passing on the trait, and so on *ad infinitum*. The reward, if not the aim, of the whole process was the survival of the fittest in this gentlemanly struggle for existence.

It is an interesting fact that the famous phrase "the survival of the fittest" was not coined by Charles Darwin; he quoted it from the philosopher Herbert Spencer—and neither of these great Victorians ever seems to have used the full expression "survival of the fittest in the struggle for existence." This catch phrase has, however, been commonly accepted as expressing the basic dogma of evolution. The trouble with this is that it might lead you to expect, as its logical extension, that the dry earth would be exploited by a single superplant. You could also be excused for wondering why there are still any cycads, let alone ferns and mosses, growing today. The ironical thing is that if survival is really the key attribute of success "in the struggle for existence," the blue-green algae must be counted among the fittest of all plants, because there is no disputing the fact that they have survived for an extremely long time. Thus there is much more to evolution than can be expressed in a catch phrase.

It has been well said that evolutionary fitness is compounded of two things: the fitness of the organism and the fitness of the environment; and it goes without saying that you can't have one

Of the many types of foul-smelling stinkhorn fungus, one of the strangest is Dictyophora. *The function of its lace petticoat is obscure, unless it is to attract naturalists.*

Wherever there is potential for life, some living thing will exploit it: solidifed lava (left) becomes the home of one plant colonizer, carpetweed, and enterprising dandelions (right) take up their abode in a heap of rusty cans and broken bottles.

without the other. Interaction between organism and environment is of great importance, but interaction between organisms is just as important, if not more so, to the process of evolution. This is because one set of organisms is often a very important part of the environment of others. To put it in its simplest terms, trees create shade and other plants can evolve to tolerate that shade. Similarly, until there were plants, the evolution of herbivorous animals was impossible. The interactions of organisms are some of the causal factors of the diversity we see around us.

Every time you go to the country and look at flowers, the facts are all around you. The flowers you see are not growing alone, but are members of more or less organized groups or communities to which we give the general name of vegetation. Most languages include special words to describe different types of vegetation—words such as grassland, forest, meadow, and swamp. Similar words are found in even the most primitive language. A good example is the language of the Bushmen, which, although it includes only three numerals (words for 1, 2, and 3), does have words for all the common types of vegetation in the area of South Africa in which they live. This is not surprising, because types of vegetation are the main signs of the Bushman's landscape; they tell him something about the potential of his environment. His survival in a given area depends on his ability to assess its potential for food, water.

and shelter, and his evolved language includes words that allow him to communicate these signs. This is all part of the package deal of evolution in that area.

Wherever you are, if you clear a piece of ground of all the plants growing on it and leave it untended, open for colonization, other plants will soon begin to make use of the potential of the plot. Very often, most of the colonists will be new plants—that is, plants that were not present in the original vegetation. If you try this experiment in your own back yard, many of the pioneers will be what we normally call weeds. I always like to keep a corner of my own garden set aside especially for weeds; this is a sort of perennial experiment, to see exactly what comes up, and there's almost always a surprise or two. It is also great fun to give the plot a bit of pre-treatment—a little fertilizer or a dose of tea-leaves, for example. All such treatments change the potential in some way, and the system responds to the change.

Don't get the idea that this is just an excuse for my not digging the garden. I have to dig it quite often, for if I didn't dig my weed patch over each year, the experiment would rapidly change. The weeds would persist only for a few years, and, as they went, grasses and shrubs would come in to swell the ranks until—at least in my garden—seedlings of ash and sycamore would make their appearance and make it felt. If I permitted the trees to get a firm foothold, the changes would slow down as the vegetation took on a specific structure and started to stabilize. In fact, the open plot would gradually become woodland. But don't be misled into thinking that this would happen only at the bottom of my garden.

Next time you are riding in a train through the outskirts of a large town or, better still, are coming in to land at one of those ill-placed international airports, take a look at all the forgotten corners in the downtown districts. Each of the little triangles and odd-shaped strips of land that have been cut off from development and left untended between walls and yards is usually overflowing with chlorophyll, green leaves of shrubs and trees reaching up to the light. Unfortunately, modern high-rise buildings in planned developments are usually furnished with wall-to-wall turf. But wherever there is some unplanned city space you will see a green oasis. And each such oasis screams the other dogma of evolution: *wherever there is potential for life, evolution will use that potential to the full.*

Any piece of terrain left untended will slowly but surely become covered with vegetation. And the process by which this happens is called *succession*. Succession can transform barren rock into living forest, or open water into forest rooted in dry soil. This process is an integral part of evolution.

A piece of bare rock may not look the most likely place on which a plant would evolve to grow—and yet this is the habitat of the minute pyrenocarpous lichens. Their simple dual tissues of algal filaments and fungal hyphae grow down into the actual substance of the rock, pitting it with life. Once this has happened, other lichens join in the simple community, until the rock surface becomes covered with a living skin of crustose lichens. If lichens could scream, it would be unbearable to sit down for a rest while out on a mountain walk, because every rock is covered with the

spreading colonies of these hardy plants. Although they cannot make their presence felt vocally, they certainly brighten up the mountain scene with their often vivid colors and patterns. And they do something else that is perhaps more significant: they play a major role in eroding the surface of the rock.

In time the surface of even the hardest rocks (such as dolerite, which is a favorite for road surfacing) becomes soft and takes on something of the consistency of wet fiberboard. The organic matter in the rock holds water, which alternately freezes and thaws during the cold of winter, disrupting the surface layers even more and allowing the lichens to penetrate more deeply. Finally, great flakes of the rock fall away from the surface. On vertical and near-vertical rock faces this flaking process—technically known as *exfoliation* —leads to a steady-state situation, because, once the soft rock layers have fallen off, new bare rock is exposed ready for recolonization. The process of succession is back to square one and so the lichens start the whole process in motion once more. But even in the early stages of succession, other plants—including the seedlings of trees— can root into the layers of rotting rock, indicating what is inevitably to follow.

If the surface of the rock is not at too great an angle, the flakes will remain in place instead of dropping off. Under these circumstances the rate of change can increase many-fold as the crustose lichens are replaced by more rapidly growing fruticose lichens, which add a considerable amount of organic matter to the developing soil. At this stage in succession the mosses make their appearance as abundant members of the changing vegetation, moving in to colonize the rock surface that has been at least partly prepared by the lichens. One of the most interesting of these newer colonists is the rock moss, which is well adapted for life on rocks. This moss has two types of *rhizoids* (minute rootlike structures)—or perhaps really only one type, which can behave in two different ways. If a rhizoid comes into contact with the edge of a small pit in the rock, it grows down into the pit and, like a rock gymnast, "clenches its fist," thereby anchoring itself firmly to the rock. The rhizoid simulates the clenched

This forest was once barren rock. Over the ages the rock has been eroded by wind, weather, and the action of plants, such as fungi and lichens; now only boulders remain—and each of them is gradually being covered by mosses and liverworts.

fist by developing a swollen pad of tissue at its tip. However, if there is no crack, the pad of tissue forms an adhesive disk on the rock surface.

The rock moss is an insignificant red-brown color and becomes readily recognizable only when it produces its somewhat peculiar spore capsules. Each capsule looks something like a Chinese lantern, with four slits running lengthwise down the fruit. When the weather is dry the slits open up, allowing the spores to be released for dispersal by the wind, and to fall on other rocks that have reached the right stage of erosion for colonization. Many other mosses now join ranks with the rock moss, so that at this stage a rock becomes covered with "cushions" of moss. Soon the entire rock surface is hidden under a thick mat of lichens, liverworts, and mosses. The mat contains minerals eroded from the rocks and other mineral matter that has been blown in and trapped among the shoots, all intermixed with the humus produced by the growing plants. It is in this cap of embryonic soil that the seeds of many other plants—herbs, shrubs, even trees—find a firm footing and a place to begin life.

You may find it almost impossible to imagine how a large piece of rock can become incorporated into the floor of a forest, disappearing to leave no trace. There is no getting away from the fact that it takes a long time, but there is also no getting away from the fact that it has happened time and time again. Boulder-strewn slopes are especially fascinating to study, because they often represent a patchwork of all the different phases of succession. The largest rocks still protrude out of the forest floor, each capped with mosses and lichens. Their sides, in an earlier stage of the process, are covered with crustose lichens, and patches of fresh rock have been exposed by exfoliation. Erosion, however, has partly filled in the narrow spaces between the boulders. The broken rock, together with organic matter, has produced a rich soil in which mature trees are already growing. The leafy canopies of these trees, in turn, are providing deep, cool shade in which the earlier stages of succession can continue uninterrupted by summer drought.

The process by which bare rock is transformed into high forest may seem remarkable enough, but the transformation that starts from the other extreme—deep water instead of hard rock—must sound like a pretty tall botanical story. The story is a true one, though: succession can gradually fill a deep lake to a point where the lake becomes a forest. To begin with, the water of any lake is a dilute solution of all the geochemicals present in the surface of the catchment area from which the water derives; thus it is an ideal place for the growth of plants. There are, however, many obstacles to plant life, not the least of which is the depth of the water; so the first large plants to colonize the lake are the maxi-, or to use the proper term, megaplankton such as duckweed and hornwort. These take over from the microscopic algae or miniplankton in the surface waters; and they are large enough to serve as substrata on which small benthic algae may ride high in the sun. The Germans have a lovely name for these hitchhikers: *Aufwuchs*. But, if you want to see them, you must stick your head under water and look at the megaplankton from underneath; every branch and leaf is edged with what looks like a transparent sheath, through which the sun glistens like gold. The "sheath" is the algae.

Around the edges of the lakes, and especially in shallow bays, other aquatic plants can grow rooted in the soft mud; their long stems and leaf stalks carry the leaves up to cover the surface, and so they get the best of both worlds. Largest among these rooted aquatics are the water lilies, which do more than their fair share of covering the surface rapidly. The odd spaces between lily pads soon become filled with the leaves of pond-weed and duckweed. The resultant living skin, which simply bursts with chlorophyll, shuts off the bulk of the light that used to penetrate down into the lake. Thus, any bottom-living plants die, and their remains, together with all the bits and pieces that fall from the chlorophyll canopy, begin gradually to fill the lake.

Wherever the water is shallow enough, other more ordinary rooted plants can begin to grow. But because there is no space left on the surface, the only plants that can succeed in this highly competitive world are those that can push themselves up above the top of the water. Bulrushes, reed mace (better known, perhaps, as cattails), reeds, and sedges grow upward and begin to take over as the process of succession speeds up. Like the water lilies, these hardy plants are perennials; each year they put on a massive amount of new growth, storing part as food for the winter in large, juicy, underground stems called *rhizomes*. These food stores allow them to "get cracking" very early in spring, and with this advantage they become so abundant that the surface floaters are gradually eradicated.

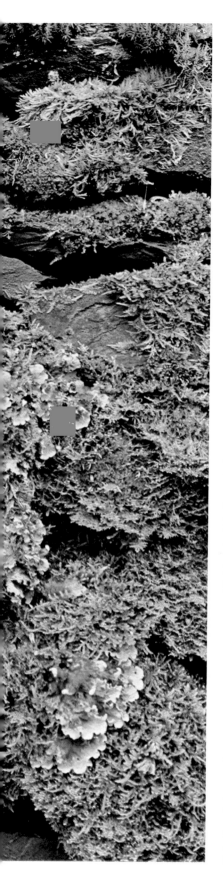

Above: an alga and a fungus together can produce delicate, yet incredibly tough, lichen bodies that can colonize even such an inhospitable habitat as a gravestone, coloring its dull surface.

Left: stone walls in England's Lake District may be proof against straying animals, but not against the invasion of lichens. This wall is the home of spreading cushions of mosses as well.

Right: given time, succession can change barren rock into fertile forest. In the early stages, lichens "eat" into the surface of the rock, producing niches in which even tree seedlings, such as the birch, can take root.

How a Lake Becomes Dry Land

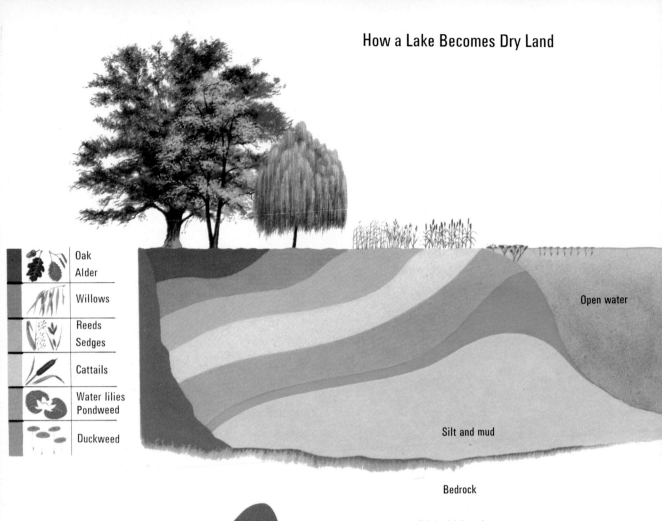

	Oak	
	Alder	
	Willows	
	Reeds	
	Sedges	
	Cattails	
	Water lilies	
	Pondweed	
	Duckweed	

Open water

Silt and mud

Bedrock

Original lake edge

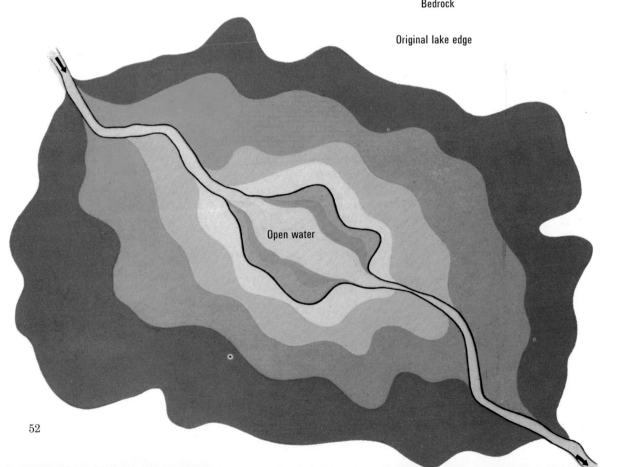

Open water

The process of succession can gradually transform a lake into a forest. As the photograph above shows, the open water is first colonized by floating plants which, together with their remains and the silt and mud brought in by a stream, begin to fill the lake. As the water becomes shallower, other types of plant can colonize the margins of the shrinking lake. In time, the mat of vegetation supported on a thick layer of peat becomes stable enough to allow the growth of trees. The zones that can be seen around the margin of the lake show the current situation in this disappearing act, but the real record of such change lies in the peat beneath the forest surface, where pollen grains and fragments of decaying plant material can be found.

Left: the illustrations show in cross section (above) and plan (below) how a lake can become dry land. As the remains of plants growing in the water and around the edge gradually fill in the basin they provide a foothold for other varieties of plants, leading eventually to the growth of trees.

Succession is now going on apace as every fall the great mat of vegetation dies. What was once open water becomes marshland, which, although wet after the winter rains, is dry enough in the summer to allow willow trees to root and flourish. Now comes the critical point in the transition: each fall, the willows cast their leaves onto the ground, and these rapidly speed up the filling-in process by building the soil up above even the winter flood level. The ground is now firm enough for alder trees to grow, lifting their gray-black trunks over 80 feet into the dry air, and they, too, add their thick green leaves to the mat of humus.

Once this mat of decaying vegetation begins to rise above the ground water level, two striking changes take place. First, the process slows down

53

because the upper layers of humus are now open to the air, which contains plenty of oxygen; and so the decomposers—bacteria and fungi—can really get down to their job of breaking down the organic matter. Secondly, an aerated soil means that many animals—worms, insects, and the like—can move in and begin to exploit the potential of the habitat; as they do so, they change its structure and help to produce a rich soil, dry enough yet not too dry for the final stage in the process. Trees such as oak and ash can now move

Below: among the first plants to colonize a lake are the floating duckweeds. They form a green skin on the water, filling the spaces between the plants that take root around the edge.

in to produce a structured forest system, which is made up of many different sorts of plants and animals, each of which finds the potential necessary for its survival. Thus in this forest—which was once the watery expanse of a lake but, through the process of succession, has become dry land—each organism has its own specific role to play in the economy of the living system.

The plankton biscuit that could have been prepared from the upper layers of the original lake would have been a fraction of an inch thick, with a total live weight of well under one ounce a square yard. This represented the total potential of the original environment to the plankton. The mixed forest is about 100 feet high and, although not

solid living material, has a live weight standing crop in the region of 20 pounds in each square yard, which is a 5000-fold increase in potential. The price paid for the modification of the environment is represented by the energy stored in the organic matter that supports the forest.

The amount of stored energy required and the length of time it takes to transform a lake into a forest depend mainly on the size and depth of the lake and, to a lesser extent, on the nutrient content of the water and the productivity of the colonizing vegetation. Some lakes are so large and deep that the process makes very little impression on them, but, nevertheless, no lake is entirely free from the relentless encroachment of vegetation.

Productivity is a measure of the rate at which new material is produced. It would be an easy mistake to equate productivity and standing crop; related they are, but the relationship is not a simple one. Standing crop is the actual weight of plant material present at any one time, and productivity is the increase in weight over a given period of time. The plankton has a very small standing crop, but the component algae have short life-spans and so the open ocean can produce up to half a pound of material per square yard

Below: this is what a body of water looks like in a late stage of colonization by invading plants. The water lilies growing in the center are surrounded by a broad band of cattails.

per year. Despite its massive standing crop, the mixed forest is only six times as productive.

The lake thus gradually shrinks and disappears, lost under the developing forest. And don't get the idea that this happens only in special places. Wherever you live, especially in the temperate regions, there will be lakes nearby undergoing the process. It can be an eerie feeling to walk through one of these developing woodlands and find that the ground gets more soggy with each step. While you are still among oak and ash trees, the going is not too bad, but by the time alder gives way to willow, it is getting much worse; and unless you know your plants very well, it would be wisest to turn back, because the route will soon be very treacherous. It is in such situations that will-o'-the-wisps are said to have led many a traveler astray—forever lost in the morass that fringes a dying lake.

Will-o'-the-wisps are not a figment of romantic imagination. They are, in fact, very real phenomena: tiny spurts of flame produced by certain types of marsh gas that ignite spontaneously when they reach the oxygen-rich air above the great sludge of partial decay. I don't know whether these flickering lights have ever led anyone to his doom. But I do know that they are absolute proof that there is no oxygen down there in the developing peat mat; and that is exactly why the lake has filled up. The lack of subsurface oxygen tells us that, in all probability, bacteria and fungi have not had a chance to break down the plant material that forms the peat, and it will therefore be in a good state of preservation.

Because the preserved plant material is a record of the process of succession, it is thus possible to dip back into history; all that is needed is an apparatus for removing a core of peat from below the forest. This done, all is revealed. There at the bottom of the core are the remains of the pondweeds and water lilies that started off the whole thing. Higher up we find, in turn, the fruits of cattail and sedge, flower scales of willow and alder, and, finally, acorns—a complete record of the process of succession, proving it has taken place right on that spot.

Unfortunately, modern farming methods have changed our landscapes so much that few of these mysterious damp woodlands are left; the dying lakes are now farmed to their living edges, because the peaty soils, once drained, are very rich. Nevertheless, before man began to play a dominant role in the landscape, the process of

56

succession gradually changed bare rock and open water into productive forest. What is more, wherever man leaves the habitat to itself, succession soon gets under way again, because, as we have seen earlier in this chapter, wherever there is potential for life, evolution will exploit that potential to the full.

My favorite time for a woodland walk is in the early winter, when there are still piles of dead leaves to be scrunched through, and it is easy to see the structure of the forest. Overhead, I can see the intricate crowns of branches that bore last summer's canopy of leaves, which shaded the forest floor, and provided the cool, damp habitat necessary for shade-tolerant plants. In this shade, plants such as delicate wood anemones and bluebells completed most of their growth in the spring and early summer, before the full thickness of the canopy developed; and they, in turn, produced more cool shade, so that close to the ground it was always wet enough for the prothalli of ferns to grow, intermixed with mosses, liverworts, and shining colonies of *Nostoc*. Very often, when the first frost is on the ground, it is possible to find the new shoots of mosses, bright green and bursting with life, making use of the winter sunlight that shines down through the bare spokes of the canopy. Below ground there is a similar stratification: the rhizoids of the mosses only just penetrate the surface humus; the rhizomes of the ferns lie deeper in the forest soil; while the roots of shrubs and trees grow deepest of all. The whole forest is an integrated, living system, which is structured in both space and time, thus making the fullest use of all the resources of the habitat—light, water, and nutrients.

Large amounts of those nutrients are present in the great standing crop of timber and herbage. All the key geochemicals—carbon, potassium, calcium, phosphorus, and nitrogen—are there, locked up and temporarily out of circulation. It is the decomposers that accomplish the essential job of releasing the nutrients ready for recirculation. Though usually out of sight in the soil, the decomposers work throughout the year; if the cycle of life were not completed by them, the whole thing would soon come to a standstill, starved of these essential nutrients.

Bluebells in full bloom carpet an English woodland in late spring and early summer. They have completed most of their growth, and built up an underground food store (bulbs) before the trees come into full leaf and shade them.

Above: a group of fungi—"kings of the refuse-disposal operatives"—at work in the leaf litter of the woodland floor. They are breaking down dead organic matter into its component elements.

Below: each one of these puffballs is filled with hundreds of thousands of spores, which when ripe will be puffed out in a cloud of rain, wind, or by being kicked by passing animals.

Among the key components of all types of vegetation are the fungi, the kings of the refuse-disposal operatives. Their hyphae ramify through everything that contains the potential of dead organic matter, breaking it down and returning the carbon to the atmosphere, as carbon dioxide, and the other nutrients to the soil, where they will be ready for re-use in the flush of the following spring. Every tree trunk, branch, twig, and leaf is scheduled for destruction. But as one of the basic laws of the universe states. "Matter can be neither created nor destroyed." And so, on their death, plants are simply reduced to their component geochemicals, and thus made ready for recycling, via the soil, to the next generation of biochemicals.

It is mainly in the fall that the fungi make their presence known—at least to woodland visitors like us—for it is then that many of them produce their fruiting bodies. But they are there throughout the year; burrowing tunnels of decay through the wood and leaves, they form a great, seething, intricate mass, which penetrates every source of energy in its path. At certain times of the year, and under certain climatic conditions, they undergo a radical change in their pattern of growth. Knots of hyphae begin to form near the surface, and these grow together to form small, dense bodies which, although composed only of interwoven threads, have enough substance to look and feel like solid tissue. The name given to a real tissue formed of living cells joined together is *parenchyma*, and so this false tissue is called *pseudoparenchyma*. The lump of pseudoparenchyma rapidly grows out toward the light, developing into a fruiting body—a parasol-like toadstool, a bracket-like bracket fungus, or one of the many other weirdly shaped fruit bodies.

Most of these conspicuous fruiting bodies will produce either delicate plates of tissue that radiate out from their stalks or ordered arrays of pores. The radiating kind (which we commonly call toadstools) are the gill-bearing fungi or *Agarics*; the others (of which bracket fungi are the best example) are called *Polypores*. Both have the same function, which is to produce many tens of thousands of spores and release them into the air. The spores are produced by, and released from, the tips of special hyphae that line the walls of both the gills and the pores. Under the force of

Left: one common name for all fungi that are shaped like umbrellas is "toadstool." Appropriately, a common toad sits near these toadstools, which are called sulfur tuft fungi.

gravity, the spores fall out into the moving air-stream above the ground, to be carried away to other hospitably rotting habitats.

The fungi with the bracket type of fruiting body must be at an advantage because, as they grow on rotten trunks, the spores have farther to fall and there is therefore more likelihood that they will be carried a considerable distance by the wind. However, there are disadvantages to this way of life because, as the wood gets more and more rotten, the branches are increasingly likely to fall off, and sometimes the entire tree trunk will topple down, bringing the fungus sharply back to ground level.

The design of the mushroom type of fruiting body is based on the necessity to raise the gills or pores as high above the surface of the ground as possible, to give the spores the best chance in life. The shapes of the fruiting bodies range from the tiny, delicate ones that look like miniature parasols on long, thin handles to those that have great, solid caps supported by fat, juicy stalks.

But although gravity and air currents play a large part in spore dispersal, not all fungi are entirely dependent on these aids. The puffballs, for example, make use of raindrops (falling rain being a form of gravity in disguise). The puffball produces its spores inside a tough but paper-thin bag, which develops a hole at the top when ripe. Raindrops falling on the outside of the bag cause it to act like a pair of bellows, puffing out the spores. One major drawback to this mechanism is that the spores are apt to be washed down into the soil by the rain before they have traveled very far. So it seems likely that, once the first spores have gone, the partly empty bag is further activated by the wind. At any rate, the spores soon disappear. Incidentally, it can be great fun making clouds of spores puff out by prodding the bellows with a finger!

Most peculiar of all fungi are the stinkhorns, which live up to their name by pervading the

Above: no fungus is more aptly named than the earth star. The outer layer of the fruit body splits open to form the starlike part, revealing and supporting the spore case, which both looks and functions like a puffball.

Right: there is nothing to make one think of the stars in this thoroughly earthbound stinkhorn. It gives off a strong aroma, most unpleasant to human nostrils but unfailingly attractive to flies, which crawl avidly over the sticky spores that cover its apex. Thus the flies become agents of dispersion, for they involuntarily carry the stinkhorn spores to new breeding grounds.

forest with a never-to-be-forgotten stench. The fruit body starts life looking like an egg, but soon cracks to reveal a rapidly lengthening *stipe*, or stalk, that bears sticky spores at its apex. Flies, attracted by the smell, crawl over the top of the stinkhorn, each inadvertently collecting some spores, which it carries to its next resting place.

One group of fungi go even further in their dependence on animals; they not only depend on them for dispersal of their spores, but they also live exclusively on animal dung. Anywhere in the world where there are dung-producing animals, there are *coprophilous* fungi that make use of them. The spores of coprophilous fungi can pass

Above: this object is no fictitious monster from a horror movie. It is, in fact, a shapeless jellylike mass—a slime mold—which lives in a world of decaying matter, "crawling" about and silently engulfing any food that lies in its path. A denizen of damp woodlands, this organism is not an animal despite its mobility, for under certain conditions it will remain stationary and produce a fruiting plantlike body containing spores.

Opposite page: the fungi shown in the picture nourish themselves exclusively on the energy present in animal dung. In a world where there are so many different varieties of fungi and bacteria, nothing can ever go to waste.

undigested through the animal's gut, to emerge ready-planted in their own dung garden. This is organic recyling *par excellence*.

Finally, there is one group of fungi that—far from being absolutely dependent on animals, like their coprophilous relatives—do not seem to have made up their minds whether they are themselves plants or animals. These are the slime molds, which form delicate, often brightly colored patterns on all sorts of decaying vegetable matter. If you have enough patience and can keep your mind on the job, you can actually watch these apparently disorganized organisms slowly move about. The body consists of nothing more than a mass of cells, which, because they lack rigid cell walls, are capable of crawling slowly. To make them seem even more like animals, they can engulf and digest organic matter that lies in their path. So why are the slime molds reckoned to be plants, when everything about them says "animal"? The answer is that not quite everything does say "animal," because the slime mold can stop its disorganized perambulations and become organized. When this happens, the cells stream together and rapidly produce a tiny fruiting body full of spores.

Mysterious as the slime molds may seem, they are quite common plants as long as you look in the right places for them. If you are lucky enough to find an amenable one, you might try adopting it as a pet. I once kept a bright yellow one for about two months; it lived on damp blotting-paper in a covered dish, and I fed it on porridge oats. It did take me numerous attempts before I found one that would live in captivity, but it was well worth the effort; and I must confess that it was a sad day when I found it had dried up. Unfortunately, it never fruited. But you may have more success. So why not look for one to take home with you, next time you go for a stroll in the woods? (You need only a very small bit.)

With the enormous number of things to look at, I never get very far into a wood before it is time to come out again. Everywhere I look there is something going on, and each something is to a large extent dependent on all the other somethings that make the woodland what it is. At first sight, a woodland community looks about as organized as the wandering slime mold. However, once you start to study even the most obvious inter-relationships, it becomes clear that the parts of the woodland are as integrated as are the parts of the most complex organism. It was Sir Arthur

Tansley, the great British plant ecologist, who first coined the term *ecosystem* to describe such a working unit of life. Included within the definition are animals as well as plants. Until there were plants there could not be herbivores; until there were herbivores the evolution of carnivores was impossible. And without decomposers, the whole system would seize up.

Interdependence is the key word. It explains the diversity we see everywhere in nature; each organism is fitted, through evolution, to do its own particular job in its own special ecosystem. What we see is not one Master Organism, but all the organisms together exploiting the environment.

63

Mixed Vegetation

Two contrasting scenes from the world of vegetation: a polar desert where permanent life is impossible because most of the abundant water is permanently frozen; and a beechwood in the temperate zone where a seasonal climate and sufficient rainfall give great potential for both plant and animal life.

Imagine yourself as the world's first botanical cosmonaut and view the planet from afar. What you will see is that the landmasses are zoned, with the boundaries of the zones running very roughly at right angles to the north/south axis of the globe. Although the zones are not clear-cut and there are major disjunctions, they are real phenomena, and not merely part of an imaginary space flight. Each zone is characterized by certain types of climax vegetation (that is, relatively

stable vegetation at the top of its ladder of succession); and, for the most part, corresponding zones are to be found in both hemispheres—although, because there is much less dry land in the temperate and subantarctic regions in the Southern Hemisphere, the zonation is less well marked "down under." Thanks to the painstaking records of earlier travelers, we know that the zonation existed long before space flights were possible. Thanks to the work of botanists, we can also estimate the potential of each zone, in terms of standing crop and productivity.

If your orbiting laboratory had a system for recording and integrating the temperature of the earth's surface, it would soon become obvious that the "hot spots," that is, the areas that receive maximum sunlight, are not the zones of maximum photosynthesis. Far from it. The main area of tropical rain forest, which occupies part of the Amazon Basin, lies in a region that receives only

about 687,000 calories of sunlight energy on every square inch of its productive surface. The hottest regions of the world receive twice as much sunlight as that—but they are desert. What the deserts lack, the tropical rain forest has in abundance: water. Clouds and rainfall go together; you can't have one without the other. And the more cloud there is, the less sunshine will get through to the ground. It's as simple as that! Life requires energy, minerals, and water. Most rock types can, in time, provide an adequate supply of mineral nutrients. But large areas of dry land lack much life because they are dry. And they are dry because of the patterns of climate, the distance from the sea, and the presence of mountains in the path of the moisture-laden onshore winds.

There is another variable factor, which is strictly related to the north-south axis: the length of the cold season and, within Arctic latitudes, the length of the polar night. At the equator, there is a perfect daily climate; the sun is "on" for 12 hours and "off" for 12 hours in each daily cycle, and the variation in temperature is little more than 36°F. In complete contrast, the sun at the poles is "on" for six whole months and then completely disappears below the horizon for the next six months. Moving either north or south from the equator, the perfect *daily* climate is gradually replaced by the perfect *seasonal* climate.

Perfect the seasonal climate may be in the eyes of the climatologists, but it is far from perfect when taken from the point of view of someone or something that must live there. It is the north-south axis that presents the two main problems to the evolution both of plants and of ecosystems. The first problem is one of energy, because any living organism must be able to store sufficient food material to keep it ticking throughout the long winter night. The second problem, although related to energy, is less direct; it's a question of water. Once the sun has been "switched off," so has the water supply. There is plenty of water still around, but it is frozen solid and of little use.

The main vegetational zones are thus related to the seasonal climates of the earth. In this chapter we shall include polar desert, tundra, taiga, temperate-zone forests, and tropical rain forest. Each vegetation type is an expression of the

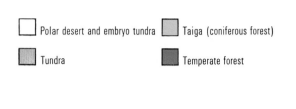

Long before satellites made space photographs possible, botanists found that vegetation grew in zones according to climate. These vegetation zones show up in the map opposite.

66

World Vegetation Zones

Grassland (seasonal)   Savanna   Desert   Mountain

Mediterranean   Semi-desert extratropical   Tropical forest   Mangrove

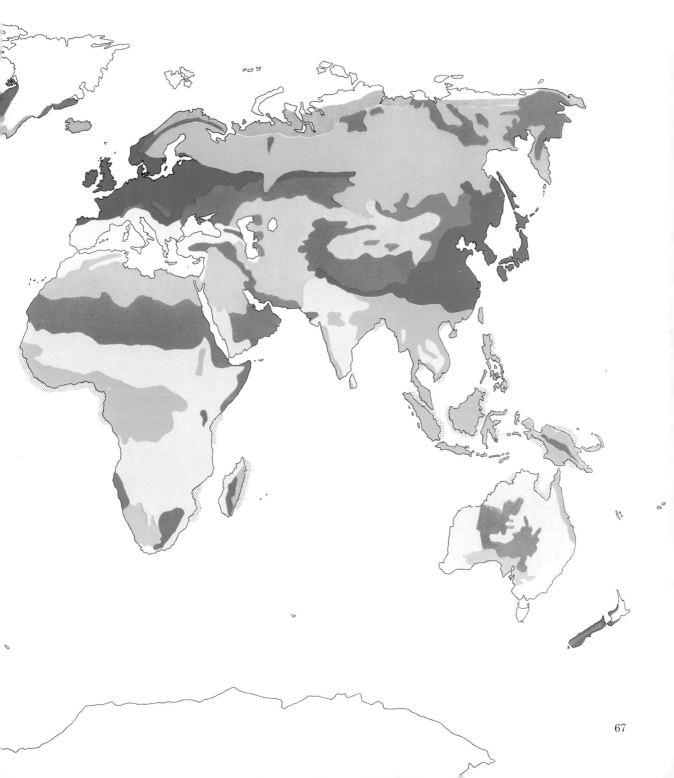

How Plant Form Varies with Environment

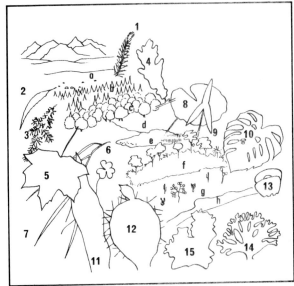

a **Tundra**
  1 Fir clubmoss
  2 Sandbar willow

b **Coniferous Forest**
  3 Western hemlock

c **Deciduous Forest**
  4 Oak
  5 Maple

d **Grassland**
  6 Clover
  7 Meadow grass

e **Lake**
  8 Cow lily 9 Arrowhead

f **Tropical Forest**
  10 Ceriman
  11 Snake-plant *(Sansevieria)*

g **Desert**
  12 Prickly pear cactus
  (Leaves absent—stem has
  taken over leaf functions)
  13 "Living stone" *(Lithops)*

h **Seashore**
  14 *Callophyllis* seaweed
  15 Sea lettuce

The evolution of true leaves opened up the various environments of earth to the plant kingdom. This illustration shows some of the leaves and other organs of photosynthesis that you might find on a journey from the cold wastes of the tundra through temperate and tropical regions to the sea. The tundra plants and conifer leaves are typically small. Combined with a thick protective cuticle, this smallness serves to reduce water loss in a part of the world where unfrozen water is in short supply. The oak and maple trees have broad leaves that make the most of spring and summer sunshine, but which are shed before the winter frosts. Low-growing grasses thrive in a wide range of climates and soils. The narrow grass leaves spring from the base of the plant and so continue to grow as the upper ends are grazed or cut off. The lake-dwelling cow lily has a broad leaf composed of air-filled cells that keep it afloat and the pores, or stomata, are concentrated on the dry upper side. In the tropical forest many leaves are large and glossy. The ceriman leaf is unique in having natural holes, the exact function of which is obscure, but it has been suggested that they prevent rain damage by allowing water to filter through. The two desert plants both show characteristic adaptations for conserving moisture; one has leaves reduced to sharp spines, and the other to fleshy, water-storing lobes that resemble stones. In shallow offshore waters the red and green seaweeds represent the early plant form in which there is no clear division into leaves and stems. The whole plant is able to photosynthesize, taking in all the gases and water that it needs over its entire surface.

"Red snow" is really a green alga that contains a red pigment. Millions of these tiny plants live among the snow crystals at temperatures that seldom rise above freezing.

current state of affairs, both of the evolution of the individual plants and of the ecosystem as a whole, which together reap the full potential of each seasonal climate.

The polar deserts are built on great reservoirs of solid water, of which only the surface layers ever melt, even though the sun may shine steadily for six months at a time. The melt waters gather minerals by solution from dust that was trapped in the ice when it formed, as well as from dust blown in during the long winter. So the potential for life is there, and one small plant has evolved to make use of it. The plant is *Chlamydomonas nivalis*, which, although a green alga, contains a red pigment, hematochrome; wherever it is present in sufficient quantity, it colors the snow red. Under even a low-power microscope, it may be

seen swimming and squeezing in a rather frenzied way between the ice crystals. It is difficult to gauge the length of the growing period of this tiny plant because, although the polar summer may last for months, the surface of the glacier is a patchwork of melt and re-freeze, and the separate mini-habitats may not last very long. The snow alga has to fit its whole life cycle into a few hectic days of summer—growth, reproduction, and preparation for the long winter ahead.

Similar environments exist at the top of all the highest mountains in the world, wherever there are glaciers and permanent snow fields. Outside the polar regions, however, they lack the perfect seasonal climate; and those that lie close to the equator experience the regular 12-hour "on/off" cycle. In this perfect daily climate every day can be a red alga day.

The word "tundra" came hurtling into the headlines a few years ago with the discovery of oil in the Arctic. Almost overnight, something that to most people conjured up a picture of harsh bleakness became something wonderful—a valuable, living system, so fragile that it must be treated with great care, lest it be destroyed forever. This view is somewhat surprising when you consider the enormous area that is covered by tundra, and the fact that it has developed in the harsh climate around the ice cap—a climate that has undergone successive and massive changes in recent millennia. Easily disturbed the tundra may be, but it should certainly not be called fragile.

In the tundra, ground temperature remains at below freezing point for 12 months of every year, and so, in place of ground water, there is ground ice! This is the great deep freeze, the *permafrost*, that paves the way to the poles. Each summer, the permafrost warms up from the top downward, and wherever enough water is present, the melt supersaturates this defrosting surface layer. It is a weird experience to walk on the melting mass; if you jump up and down in one place, the ground takes on a semifluid consistency, and the effect spreads until quite large stones several feet away from where you are jumping begin to move.

In a similar way, the pressures that are set up in this semiliquid active layer during the cycles of freeze and thaw move and sort the stones and sediments into remarkably regular patterns, which look almost man-made, especially when viewed from the air. And it is in the depressions among these ice-worked patterns that the first sign of vegetation can be seen. Living on the well-watered substrate, which consists of rocks that have been split and ground by ice action, the plants have a ready supply of minerals.

The first plants to play the game of community living in these depressions between the groups of stones and sediments are blue-green algae, lichens, and mosses; but they are soon joined by a number of flowering plants, including the world's hardiest tree, the Arctic willow. It is this diminutive deciduous tree that most obviously reflects the seasons of the tundra. At the onset of spring, the leaves open quickly, together with the catkins, which are surprisingly large for such a small tree. In early summer, the scene turns prematurely white as a result of the fluffy lint on the willow fruits. But soon the yellows and reds of fall vie with the colors of the sun as it sets.

Spring in the tundra is a riot of color, with yellow poppies, blue lupins, pink bistort, and flame rhododendrons bursting into flower. And in the fall there is plenty of fruit: bearberries, crowberries, bilberries, and—most succulent of all—the yellow-red Arctic cloudberries, which the birds stock up on before their long flight

Permafrost ground in the tundra of northwestern Greenland showing the characteristic ice-worked patterns of the surface.

In depressions between the ice-worked polygons, peat-forming plant communities can thrive during the short tundra summer.

south for the winter. But because the plants, unlike animals, cannot migrate, they must stay rooted to the spot, which soon gets very cold. And during the long winter, the problem of water becomes acute, because the supply is now solid. In those areas of the tundra that get plenty of snow, the problem is partly solved, because the vegetation is protected from the desiccating winds and extreme cold by a blanket of snow. Over much of the vast world acreage of tundra, however, there is insufficient snow, and the plants are exposed to the vagaries of the full polar climate. They survive either because they are deciduous or because they are *xeromorphic*, that is, their leaves are modified to minimize water loss. The modification has been brought about through a reduction in the surface area of the leaves and through the repositioning of the all-important stomata into pits or grooves, which are filled with a weft of hairs that help to prevent the escape of moisture.

The problem of the growth of real trees in the tundra is a fascinating one. Certainly the depth and the instability of the active layer go against the possibility, because a large tree would not be able to get a firm foothold. There is also the problem of storing enough food to keep such a massive plant body "ticking over" through the long, cold winter, for once the summer sun has disappeared below the horizon the whole living system must rely on the energy stored during the short summer.

There is no getting away from the fact that it's an ice-hard life in the tundra. But the plants that can put up with it do reap the benefits of the perfect seasonal climate.

One edition of the New York Times takes the total production of—how many acres of forest is it? It would be contradictory while writing a book to suggest that not all of this verbiage is really necessary, so I shall content myself by stating that we should concentrate our efforts on recycling. The basic raw material of the paper industry is wood pulp, and the majority of the world's wood pulp has come from what once seemed to be endless forests that spanned the northern part of the globe, roughly below latitude 55° North. So much of that reserve has already gone that now even the most stunted trees growing on the verge of the tundra, which are of no use to the timber trade, are being used to make tolerable pulp.

The Russian name of "taiga" has been given to the broad belt of coniferous forest that lies between the tundra and the more diverse mixed woodlands farther south. This is doubtless because vast areas of the taiga zone are in Russia and Siberia. To those of us who have seen only man-made plantations of conifers—spruces and pines massed together in ordered array—the pattern and diversity of the natural taiga of Siberia or Canada would be difficult to imagine. The somber greens of spruce and pine are relieved by patches where larches hold up their delicate bunches of deciduous needles, allowing more light to filter through to the forest floor each spring. The trees in the natural forest are well spaced; and although the majority of them are evergreen, enough light penetrates the canopy to support a moderately lush ground flora, rich in liverworts and mosses. The wetter bottom lands, where small streams flow through the terrain, are marked out by ribbons of willow and black poplar. It is here that beavers work away at damming the streams, and the back-up waters flood quite large tracts of forest, producing con-

ditions in which swampland and peat bogs develop. The bogs of the North American taiga are called *muskegs*, from an Indian word meaning "wet, grassy place." Swamps and bogs, whether natural or beaver-made, are among the many manifestations of the process of succession filling up bodies of water.

Within the taiga zone, however, the coniferous forest reigns supreme; and when any body of water fills with peat, the trees soon follow on. Because of the waterlogged habitat, they can grow only very slowly, and so they develop a crown of many short branches, borne on a long almost bare terminal shoot. Such crowns of branches, known as "mop tops," are typical of the muskeg forest.

Each of the countless millions of trees in the taiga is made up of millions of tubular cells, which form the support/transport tissue that we

The world's hardiest tree, the arctic or dwarf willow, is a creeping deciduous shrub with outsize catkins. Barely four inches tall, the tiny tree is scaled down for survival in the harsh tundra.

Coniferous forest directly south of the tundra is known as taiga. It is heavily exploited for making wood pulp for paper.

looked at in Chapter 3. However, the water transport system of coniferous wood consists of only one type of tubular cell—the *tracheids*—which are long, very narrow cells with end walls that provide some resistance to the flow of water. In contrast, the wood of broad-leaved, flower-bearing trees, such as those found in the temperate zone immediately south of the taiga, contains two types of conducting elements: not only the long, thin tracheids but also shorter, fatter cells called *vessels*, whose end walls are perforated by large holes, thus providing much less resistance to water flow. The vessels are the peak of evolutionary plumbing equipment, at least in the plant world. Perhaps you are wondering why plants have tracheids when the vessels do such an efficient job. Unfortunately we do not know the answer to this, but it may well be that the role of the tracheids is mainly one of support.

For a long while one of the great mysteries of the plant sciences was the question of how sufficient water gets to the top of a tall tree; many

A single day's production of city newspapers uses up many precious acres of the world's ever-dwindling forest reserves.

botanists searched for hidden pumps, but to no avail. Today it is generally accepted that the mechanism is purely physical. One of the key properties of water is that it exhibits high cohesion: that is, the water molecules are bound to one another by strong forces of attraction. If you could look at a living tree trunk with some special X-ray machine, you would see that the vessels and tracheids contain continuous chains of water molecules. Wherever water is lost from the leaves by evaporation, the next water molecule moves up to take its place, dragging the rest behind it. As long as there is a supply at the bottom and loss at the top, the process will keep going; there is no need for pumps, and thus although the living cells that surround the vessels and tracheids probably do help the process, they are not its major components.

The mixed forest of the temperate zone takes over, almost imperceptibly, from the coniferous taiga. Here a wealth of tree species is found, including

red oak, maple, basswood, walnut, and linden, to name but a few. All of these add to the variety of the structured forest. Their presence is very obvious, especially toward the end of the year when, in the fall, the chlorophyll begins to break down, leaving the dying leaves suffused with some of nature's most beautiful pigments.

It is the change from green to the reds, yellows, and browns of the fall season that sets the picture of the annual cycle of growth and dormancy that governs the life of everything in this zone. The gradual loss of chlorophyll marks the end of a season's hard photosynthesizing by the broad leaves; and during that time enough food must be stored in the roots, branches, and trunks to keep the large trees well supplied throughout the winter. The various pigments are all there in the leaves throughout the year, but only when the chlorophyll begins to break down are they unmasked to show through in all their glory. The yellow, red, and brown pigments are also of importance to the process of photosynthesis, for they absorb light in other parts of the visible range. They can channel this energy, which cannot be used directly by the green pigment, into the very complex process of photosynthesis.

The maples, which exhibit some of the most beautiful fall colors, also have the most beautifully arranged leaf mosaics. Each pair of leaves is held at a specific angle to the next pair, so that as little as possible of each leaf blade is shaded by the one immediately above it. This helps to ensure that the greatest number of pigment molecules are exposed to the greatest amount of sunlight. The whole arrangement is a highly efficient light trap, and efficient it must be, because, during the summer, the tree must be able to fix enough energy for growth and reproduction, and for use during the ensuing winter. Most important of all, there must be enough stored energy to enable the buds to burst and the young leaves to unfold in the following spring, before they can begin to make their own food. The spring is, therefore, the time when a strong flow of sugar-rich sap is passing up the trunk to the leaves. In maple trees this sap is collected by man and concentrated to produce the rich maple syrup that makes a real Canadian breakfast.

The food made in the leaves of trees during the summer is stored in all the perennial parts of the tree. Although the cells that actually conduct water are dead, the woody tissues are permeated

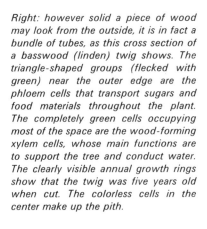

Left: fall in the mixed forests along the banks of Snake River in northwestern United States. Now as the life of the leaves comes to an end, the chlorophyll breaks down, and some of nature's loveliest colors show through in all their glory—a sight common in the temperate zone where the climatic conditions are extremely favorable to many varieties of deciduous trees.

Right: however solid a piece of wood may look from the outside, it is in fact a bundle of tubes, as this cross section of a basswood (linden) twig shows. The triangle-shaped groups (flecked with green) near the outer edge are the phloem cells that transport sugars and food materials throughout the plant. The completely green cells occupying most of the space are the wood-forming xylem cells, whose main functions are to support the tree and conduct water. The clearly visible annual growth rings show that the twig was five years old when cut. The colorless cells in the center make up the pith.

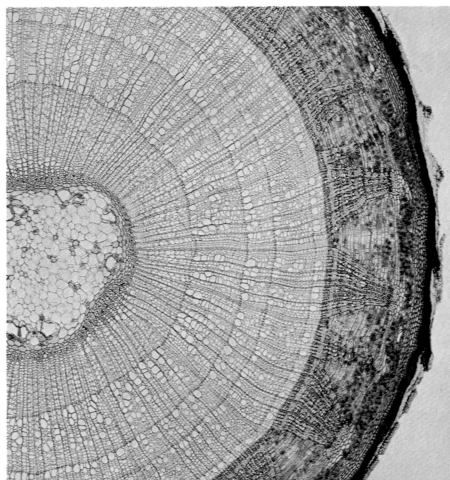

with other cells that function as living larders. It would be a nice piece of functional design to have both water and food carried round the plant in the same system of pipes, but, unfortunately, there are two reasons why this is not possible. Water flows through the plant on a one-way ticket, it comes in at the bottom and goes out at the top. There is no need to direct or control the flow, for the only thing that must be controllable is the rate of loss of water vapor from the leaves, and the stomata take care of that. When it comes to the sugars, however, both their direction of transport and their rate of flow must be controll-able. For example, at certain times of the year, foodstuffs must be translocated to the developing fruits or, in a plant such as the potato, down to the underground storage organs. This cannot be a haphazard process, but must be carefully controlled and directed. Therefore, a special tissue has evolved to do the job, and this is known as the *bast* or *phloem*. Phloem is the delicate white tissue that is found just below the dead outer cells of the bark. It is a totally living tissue, which encircles the xylem.

The cells that conduct sugars in a tree are elongated and joined end to end to form a living

In springtime the energy stores in trees are mobilized so that they may be used for new growth, especially for the development of young leaves. Sap rises from the roots to the topmost branches within the phloem tissue, or bast (which appears as the darker orange areas in the stained section of a butternut tree shown above). The photograph of a sugar maple at left shows how man makes use of this natural process. Some of the sap is diverted by means of metal tubes into the containers clustered around the base of the trunk. This sap is then concentrated, by boiling, into a sweet syrup. The best time for the collection of sap is during the early spring, when nighttime temperatures are below freezing and the days are relatively mild.

tube. The end walls that separate each member of a vertical row are perforated by relatively large pores, so that, in surface view, the cells look like sieves—and this is the reason why they are called *sieve tubes*. Sieve tubes lack one constant feature of all other living, *eucaryotic* cells (cells in which the nuclear material is enclosed in a distinct membrane), and that is a nucleus. However, the nucleus that in all probability helps control the life functions of the sieve tube, is found in the *companion cell*, which adjoins the sieve tube itself. The sieve tubes and companion cells lie packed in among other living cells that show no visible modification, and these unspecialized cells are called simply *phloem parenchyma*. The exact mechanism of transport through the sieve system is still somewhat obscure. Various theories come and go—and we must be thankful that the phloem takes no notice and gets on with its job. Perhaps most puzzling of all the facts so far discovered is that it is evidently possible to have material traveling in both directions in one sieve tube at the same time. No one has yet explained how plants manage such a complicated job of traffic control.

At the onset of autumn, the rate of sugar flow falls, and the pores in the sieve plates get blocked with a substance called *callose*, which shuts the transport system down for the winter. As the flow of sap slows down and the leaves begin to turn color, the key change that sets the temperate forest apart from others takes place. Each leaf stalk is loosened by a process called *abscission*. The living cells, including the phloem, begin to disintegrate along a special line across the base of the leaf stalk, leaving the leaf attached to the twig only by the dead woody elements that are themselves already blocked off and useless. A gust of wind snaps the final connection, and the leaf falls. Each snap-off point is, in fact, a self-inflicted wound, which could be a point of entry for fungi and bacteria. So, before the leaf falls, the wound is sealed off by the growth of new tissue just below the actual level of separation.

It is not the process of abscission itself that is unique to these forests, for all perennial plants cast their leaves at one time or another. The unique thing is the annual cycle of abscission. It is this cycle that helps control the whole life of the deciduous forest. Every year, year after year, minerals are recycled from the trees back to the soil, together with a fresh load of organic manure. Thus the whole system is kept in productive working order. Moreover, the regular cycle of

The lady's slipper orchid is among the showiest of the flowers growing in forests of the temperate zone. The lower petals are developed into an inflated slipperlike sac.

abscission opens up the ground surface to light, allowing a rich ground flora to develop in the spring and early summer.

Probably the first western description of tropical rain forest was written in 1493, by Christopher Columbus. Of the Isle of Española in the West Indies—now called Hispaniola—he said: "Its lands are . . . filled with trees of a thousand kinds and tall, and they seem to touch the sky. And I am told that they never lose their foliage, as I can understand, for I saw them as green and as lovely as they are in Spain in May and some of them were flowering, some bearing fruit and some in another stage, according to their nature."

During the following centuries there were many who visited these forests, including an increasing number of naturalists. They could hardly help noticing that life out there was not just a hotter, more humid, accelerated version of the temperate forest life they knew back home. The British naturalist Henry Bates expressed it very clearly: "In England, a woodland scene has its spring, its summer, its autumnal and its winter aspects. In the equatorial forests the aspect is the same or nearly so, every day of the year; budding, flowering, fruiting and leaf shedding are always going on in one or another species."

The key factor that controls the distribution of the rain forest is rainfall. Wherever the climatic cycle is interrupted by a dry period, the forest becomes deciduous and its whole character changes. In essence, it becomes more like a temperate forest, in which the seasons are marked by changes in the whole pace of life.

There is little doubt that the size of tropical rain forest trees has been exaggerated by some writers. The typical height of the canopy is in fact only about 130 to 200 feet, and the trunks of most of the trees are not more than three feet in girth. There are, however, authentic records of trees up to 300 feet tall and of trunks with a girth of 60 feet. Perhaps one should not blame the tellers of tall stories, because inside the forest it is very difficult to judge the exact position of the canopy. One is reminded of the old saying, "You can't see the wood for the trees," but within the forest it is a case of, "You can't see the trees for the wood." From the forest floor only the trunks are visible, and most of them look very similar because nearly all are covered in smooth bark. The leaves, flowers, and fruits are out of reach way above your head, and this presents some extremely

Above: surrounded by curl-crested toucans in South America's rain forests, the great English naturalist Henry Walter Bates (1825–92) carries on in his exploration of the basin of the River Amazon.

Left: tropical rain forest In the mountains of Tanzania, in east-central Africa. Lush vegetation such as this is evolution's usual response to the enormous potential for life that exists in the tropical latitudes wherever there is excessive rainfall all the year round.

81

awkward problems to would-be forest ecologists.

In order to study the structure and makeup of the forest, there are four courses open to you. You can learn to climb trees, or learn to aim a gun vertically upward so accurately as to hit and thus break off the right twig on the right tree, or get down to the hard work of cutting a swathe through the forest. This third method, although effective, is hardly likely to win you an award for conservation, because it requires that all the trees along a selected line must be felled. It is easy to see that any of these methods allows only a minute piece of the forest to be seen, and this could give a very misleading picture of the make-up of the whole thing. The fourth way is the best: to work alongside the foresters in a logging camp and collect the data as they obtain their wood.

Even then, when you have the leaves in your hand, your problems are only just beginning. The majority of leaves look remarkably alike—each about the size and shape of a laurel leaf, with a long blunt tip into which the nerve, or mid rib, runs (and from which the rain falls with monotonous accuracy straight down the neck of any ecologists working on the forest floor). These tips have been aptly called "drip tips." Although their exact function is still a bit of a mystery, they must have something to do with ridding the surfaces of the leaves of excess rainwater as rapidly as possible. Only if you find the flowers and fruits of a tree can you be certain which species of tree it is, and as there are no seasons in the forest this can be very difficult.

One great group of flowering plants that disobeys the "rule of the uniform leaf" is the Mimosoideae, which are in the main the tropical representatives of the worldwide pea family, the Leguminosae. One of the characteristics of this family is *compound* leaves—that is, leaves that consist of a central, often branched leafstalk bearing many small leaflets.

It is difficult for a botanist who has studied his subject in the temperate regions to imagine beans growing on plants the size of trees. But if he visits a rain forest, he will soon see these tree-sized legumes. As the British naturalist Richard Spruce wrote: "Nearly every natural order of plants has

trees amongst its representatives, grasses 60 feet or more high, violets of the size of apple trees and daisies borne on trees the size of alders."

In these most favored regions there is always plenty of rain, and the tropical sun shines for approximately 12 of every 24 hours throughout the unchanging year. So the main problem in making the most of this choice environment is the struggle up toward the light. There are three basic ways for a plant to get up and keep up at the top of this competitive forest world. One is to gain support by means of a wooden trunk—that is, to become a tree. The second is to become a climber, rooted in the soil but using a "host" tree for support. And

Below: the leaves of this hardwood-tree seedling end in long points that are aptly named drip tips, since their function is probably to shed excess water from the leaf surface.

Left: in the ideal growing conditions of the tropics, plants whose temperate-region counterparts grow to a modest size can reach gigantic proportions. The fruits of the South American flame of the forest, shown in the picture, look like enormous pea pods and tell us at once that it is a member of the pea family.

the third is to become an *epiphyte*—a plant that is non-parasitic and merely uses the "host" tree as a place on which to grow in the sunlight, high above the dark forest floor. Plants that do none of these things have, in general, relinquished direct dependence on light energy and become parasites or saprophytes. The forest plants all belong in one or other of these ecological groups.

The "support-yourself brigade" needs little introduction unless you live in the tundra. One tree trunk is very much like another. But, even in the uniform woody world, there is a certain amount of diversity, and this can be a useful pointer to the identification of the rain forest giants. For instance, the bark of many of the trees is smooth and somewhat featureless, but once it is cut, the color of the slash, both fresh and after drying, and the type, color, and smell of any exudations are often highly characteristic and can be used for identification. But this technique

The Ecological Classification of Rain Forest Plants

Plants

Self supporting both mechanically and nutritionally

Something to feed from

Trees

Herbs

Shrubs

Parasites

e.g. Rafflesia

has its hazards: for instance, certain families, such as the spurge family, exude copious quantities of latex, and to be covered in this sticky fluid adds considerably to the problems of working in the insect-laden air. The exudations may also be very poisonous—a fact well known to the tribes that hunt with poisoned blow darts.

The boles of many of the rain forest trees are not cylindrical, but fluted, the main root axes being extended back up the tree to form great buttresses that give added support. These buttress roots present problems to the forester, and a tree is often felled from a platform above the region of the roots, the choicest buttresses being cut down afterward. Other trees derive lateral support from aerial roots, which grow out from the trunk often high above the ground and grow down to root in the soil. These roots thus prop up the tree in a manner not unlike those "flying buttresses" that support the walls of certain cathedrals.

In tropical rain forest there is a continual struggle to exploit the potential of so choice an environment. In order to succeed in the quest for sunlight, plants must either be self-supporting, or they may make use of self-supporters to lean on and find a place in the sunlit canopy. Alternatively, some plants live in the dark depths of the forest and get their food from other plants. Of these, some live parasitically on growing plants, while others live as saprophytes, feeding on dead and decaying matter.

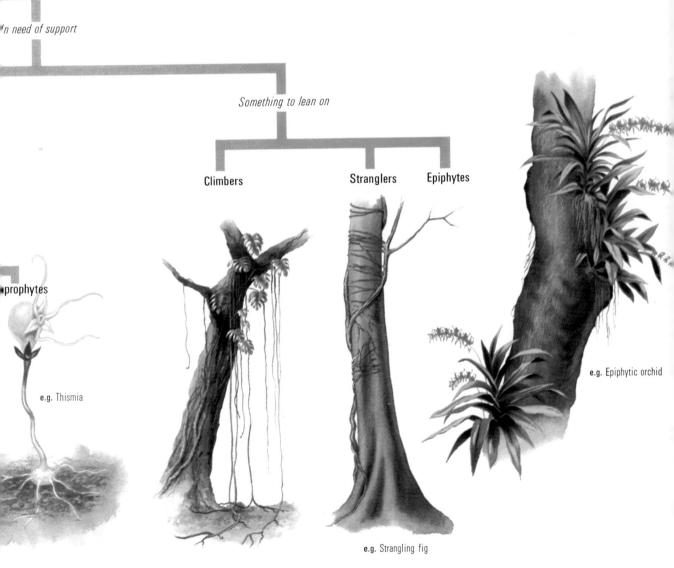

In need of support

Something to lean on

Climbers

Stranglers

Epiphytes

saprophytes

e.g. Thismia

e.g. Strangling fig

e.g. Epiphytic orchid

The amount of extra support a tree has is related, at least partly, to the type of substrate. The forests with the most bizarre root systems are those on low-lying wetlands, especially in estuaries; and this is where the mangroves reign supreme. Walking through a mangrove forest is a lesson in both balance and plant anatomy. The problem is, first, to keep your balance while walking on the mass of aerial roots, and then to try to distinguish between roots and stems. It is not too difficult, however, to identify the aerial roots while they are still growing down through the air, as the tip

of each is covered and protected by a root cap, looking like a brown thimble, which can be pulled off with ease.

Two other bizarre types of root—called "breathing" and "knee" roots—abound in the tropical swamp forest; their peculiarities are related to the fact that the sticky silts and muds are almost devoid of oxygen. Breathing roots, or *pneumatophores*, are unusual in that although they are simply side branches of the main roots, they disobey all the rules of root behavior and grow vertically upward into the air. Each breathing

Above: the tremendous weight of this Sang dragon tree in Southeast Asia is supported by great outgrowths called buttress roots. These structures are only found on trees in the tropics.

Right: the latania palm grows in swampy soils and gains support from the complex system of prop roots that grow out from its slender trunk in great profusion.

root is well supplied with *lenticels*—special tissue that has intercellular spaces—through which gaseous exchange takes place. The knee roots, on the other hand, only half disobey the rules. They grow up out of the mud, and then, under the force of gravity, turn back on themselves, growing down into the mud. At the turning point a structure that looks something like a knobbly knee is produced, and it is through this that gas exchange takes place, keeping the subterranean parts supplied with oxygen.

Strange as the root systems of the self-supporters may be, it is the "hangers-on"— the climbers, stranglers, and epiphytes—that give the equatorial forest its true character. The climbers, or *lianas*, illustrate a number of interesting problems, notable among which is how they get their leafy branches up into the canopy. The only part of the lianas visible from ground level is the tangled masses of their stems, which trail down to the forest floor where they root, often at a considerable distance from the support tree. As there are obvious problems relating to the growth of plants on the dark forest floor, it is probable that many of them must grow up along with the trees. Once in the sunlit canopy, however, they have the freedom of the forest, growing from one crown to another. Stranglers, on the other hand, start life on the branches of trees and then send down great festoons of aerial roots that can envelop the trunk to produce a living cage in which the tree is often choked to death.

The other group of hangers-on that live high on the rain forest trees in order to get a place in the sun is the epiphytes. Hardly an inch of sunlit space goes to waste; lichens and mosses festoon the twigs and branches, algae crowd the water-tracks down the tree trunks, and *epiphylls*— diminutive, leafy liverworts—form intricate patterns over the surfaces of leaves. The epiphylls are particularly well adapted for life in the forest canopy because they live on the leaves themselves and therefore intercept the light before it reaches the leaves of the host tree. Among the larger epiphytes, gigantic ferns jostle for space alongside large-leaved bromeliads and brightly colored epiphytic orchids.

Without doubt the most striking in appearance of all epiphytes are the orchids. Not all orchids are

The subterranean parts of mangrove trees, which live in oxygen-poor mud, are supplied with oxygen by aerial roots with pores.

epiphytic, of course, but some of those in the rain forest are highly modified for life out on a limb. They produce very long aerial roots that grow down into the humid air of the forest. The aerial roots are very easy to see because each of them is covered with a white sleeve, called the *velamen*, which looks somewhat like a plaster cast. Younger root tissues project from the end of the velamen, which are often green and capable of photosynthesis; and as in all roots, the tip of each is covered by a protective cap. Because of its rather spongelike structure, the velamen was for a long time thought to be able to take up water from the saturated air, thus helping to keep the plant supplied. But recent experiments have shown that it actually works in the opposite way, in that it protects the root from water loss. The only place at which water enters the plant body is through the roots, where they are appressed to the surface of the tree on which the orchid is growing. At the points of contact the roots do not develop a velamen. The question that must now be asked is: Why do these orchids produce any roots at all, as they neither support the plant nor take in water over most of their surface? But this is a question to which we have as yet no answer.

By definition, epiphytes gain only support from their hosts, whereas parasites gain both support

Left: a screw pine with long aerial support roots. Such roots, growing down into the ground from considerable heights, may look like stems.

Right: in this close-up photograph of a knee root, we get a clear view of the root's "air-breathing" pores, or lenticels. The white areas are masses of lenticels through which a supply of oxygen-rich air diffuses into the underground portions of this swamp-dwelling plant.

Below: some climbing plants, known as stranglers, may put down such quantities of aerial roots that they eventually form a living coffin for the tree that originally gave them support.

and sustenance. But no classification is perfect. Certain plants fall between the two definitions, and among these are the many tropical members of the mistletoe family. The roots of the mistletoes grow into the tissues of the host plant, from which they evidently obtain both water and minerals. We do not, however, have conclusive experimental evidence that there is also a transference of the host plant's sugars. So it is probably best to think of the mistletoes as semi-parasites— especially when we take into account the fact that they are photosynthetic.

The rule of the tropical rain forest is that wherever there is sunlight, there is chlorophyll to trap it. Thus the key to the struggle for existence is the struggle for light. Many of the leaves of the canopy trees are modified for this struggle by possessing hinges called *pulvini* at the base of the leaf stalks. Motor cells in the pulvini turn the leaf blades so that they are held at right angles to the light. This is especially important low down in the canopy, where the light is a pattern of sunflecks that move with the sun.

Below the canopy there is little light, and it is for this reason that the equatorial forests lack the lush ground layer found in broad-leaved forests of the temperate zone. Usually the only plants that can exist at ground level are saprophytes, which lack chlorophyll and live on the decaying remains of other plants, and parasites, which gain sustenance from the roots of their neighbors. Most famous of such parasites is without doubt *Rafflesia*, which boasts the largest flower in the world (up to three feet in diameter). Unfortunately, this flower, which has neither stem nor petals and which simply swells up on the stem or root of another plant, is both unattractive and distinctly malodorous.

The darkness of the forest floor presents a problem not only to the would-be ground flora but also to the canopy plants. The problem is, how can their young grow in the deep shade? This problem is in part overcome by the production of enormous seeds, which contain a large amount of food material to nourish the young plants through the process of germination. But because the seeds are heavy, they fall straight to the ground, so the saplings must grow—or try to grow—right in the shadow of their parents. The forest floor is thus often littered with fruits in various stages of growth and decay, and with many saplings each with a tuft of expanded leaves at its tip. These saplings appear to exist in a state of suspended animation until one of the canopy trees dies and a gap opens up, letting light through. The race for survival is then on, and all the saplings that are capable of growth shoot up toward the life-giving patch of clear sky. It is a race that only one can win. For the rest, the reclosing of the canopy means death. But nothing goes to waste in the living system; as the overshadowed saplings die and then decay, the minerals they contain are returned to the soil.

Above: there are two kinds of plant clinging to this tree in the Australian rain forest: an intricately patterned climber leads up the trunk to a large growth of bird's-nest fern that flourishes in the shady region just beneath the sunlit canopy.

Special Environments

The main zones of vegetation, as set out in Chapter 5, represent experiments in living in the grand manner—experiments in relation to the world's major climatic types, evolved in relation to the limiting factors of water and energy supply. It is perhaps not surprising that similar kinds of vegetation have evolved in corresponding zones in various parts of the world. The remarkable thing is that although the same types of vegetation have developed, different plants have played the leading roles in different places. In other words, the same working structures have been built with different bricks (although we must not forget that all the bricks are made out of the same raw materials, the living biochemicals).

Perhaps the most impressive case is that of the beeches. Beeches of the Northern Hemisphere produce such magnificent woodlands as those immortalized by the 18th-century English naturalist Gilbert White in his famous descriptions of the towering trees near his native village of Selborne, in Hampshire. Beeches of the Southern Hemisphere belong to a different genus, as you can tell from a quick comparison of their leaves and flowers. But a whole forest of them in a similarly placed zone south of the equator has not only the same look but the same general pattern of life as those up north. Gilbert White would not have felt too far away from home in one of the beechwoods of New Zealand.

One reason for the difference between the flora of north and south, or east and west, is that evolution of the plant kingdom did not start in one place only and then continue in a linear progression of change from its single point of origin. Evolution has proceeded—and is proceeding—the world over, at varying rates; and different groups have commenced their evolution at different places. It is therefore surprising that, time and time again, when it comes to function, evolution in widely separated regions has produced very similar end results in terms of both organisms and, especially, ecosystems. Why? It seems as if

Special climatic conditions produce special types of appropriate vegetation. In the hot dry climate of southern Spain, the colorful plants that appear with the rains of springtime are soon replaced by the parched scrublands and fields of summer.

Comparison of the Amount of Plant Material present in some Vegetation Zones

Vegetation on the earth occurs in broad zones that correspond roughly to the different climatic zones. Where the climate is warm and moist, the vegetation grows more thickly than it does in colder regions. Botanists measure the amount of vegetation that can grow in each zone in terms of the weight of plant material present at any one time, called the standing crop. *This diagram compares the standing crop of some of the major world vegetation zones.*

$\eta \longleftarrow$

Standing crop expressed as:	Polar desert	Tundra	Taiga (Coniferous forest)	Cool mixed forest	Cold deciduous forest
Grams of carbon per square meter:	Very low	50	5,500	8,000	12,500
Pounds of carbon per square yard:		0.09	10.14	14.75	23.05

Temperate forest

evolution is limited in some way and can produce only a certain number of basic structures of success—and we find these in more or less regularly repeating patterns.

It is nevertheless a fact that in certain places evolution has produced some unique plants, the like of which are not found elsewhere. In some cases they are confined to the area in which they flourish by geographical barriers; in others, because of special conditions that prevail in their native habitat. In their efforts to learn as much as possible about the whys and hows of evolution, biologists have inevitably, like Darwin, paid much attention to islands where, in isolation, evolution "got cracking" with limited raw materials, and filled all the available niches by *adaptive radiation.* Australia is an island of continental dimensions and offers an enormous range of habitats that plants can adapt to and colonize (which is just what is meant by "adaptive radiation"). Among its many unique types of organism, there is none more uniquely Australian than the eucalypts, or gum trees. The eucalypts have become adapted to just about every form of Australian forest life; they range from the gigantic trees of the coastal plain, through desert scrub, to the snow gums, which grow high up in the alpine woodlands of the Snowy Mountains. So the eucalypts are a classic example of adaptive radiation. But until fairly recent times they were to be found only in Australia.

The evolution of the eucalypts, in other words, was blocked by the fact of Australia's geo-

graphical isolation. The gum tree could get no further until man took it to other continents. Once man intervened, the picture changed. The gums are now a familiar sight the world over. They produce quick cover, a good return of timber, and that unmistakable aroma.

The Coco de Mer, or double coconut, is the most peculiar and extreme example of a plant whose dispersion was strictly limited by its place of origin. This quite ordinary-looking palm tree, with its extraordinary fruit, is confined to one area on Praslin, which is one of the Seychelles Islands, situated in the Indian Ocean. The great double coconuts were, in fact, known long before the island where they grow was discovered, for they are capable of floating in the sea and had been found washed up at various places. They were, indeed, thought to be the fruit of some submarine plant and were held in high esteem, fetching large prices at auctions because of their rarity and novelty. The first ship to visit the Seychelles, in 1609, knocked the bottom out of both the myth of their origin and their market. They take more than 10 years to mature, after which many are swept to the coast and out to sea by a river. The reason they never become established on the shores where they are stranded is that they are inland plants of a wet-forest habitat, not plants of the coast. The ordinary coconut also gets washed about by the sea, but because it can germinate and grow on coastal sands, a firm landfall may eventually become a coconut grove. So the ordinary coconut has

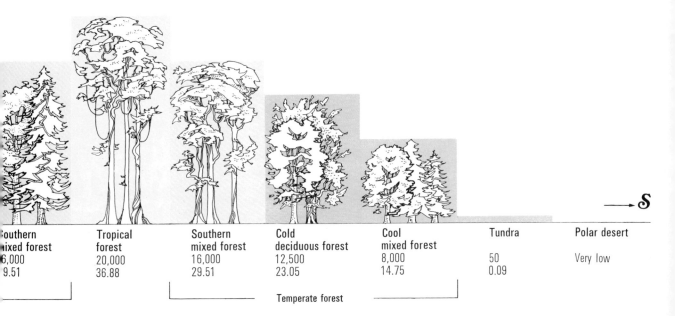

Southern mixed forest	Tropical forest	Southern mixed forest	Cold deciduous forest	Cool mixed forest	Tundra	Polar desert
6,000	20,000	16,000	12,500	8,000	50	Very low
9.51	36.88	29.51	23.05	14.75	0.09	

Temperate forest

spread throughout the tropics, in contrast to the Coco de Mer, which has presumably stuck exactly where it evolved. It is, incidentally, inedible.

Just as evolution has produced some special plants, so has it produced some special types of vegetation. This does not necessarily mean that these special types are restricted to one small area, although some of them in fact are. It means that they have developed in relation to special conditions that may be found in many different regions of the world. Thus, wherever there is insufficient water, regardless of the world zone, desert conditions will prevail; deserts are found in both the tropics and the subarctic regions. The same goes for the saline conditions found between the tides on all tidal coastlines; here the habitat conditions determine the type of vegetation, no matter what the latitude. Plants in such ecosystems are not common elsewhere; they can be considered as types of vegetation whose main characteristics have developed in relation to some limiting factor of the habitat that overrides, at least in part, the effects of the regional climate. Such habitats, which account for all the major disjunctions on the map of world vegetation, include dry deserts, saline deserts, summer-dry scrub forests (also known as Mediterranean vegetation), grassland, and peatland. Now let us look at the plant life in these special ecosystems.

Grassland, natural or man-made, is one type of ecosystem familiar to most of us. Next time you decide to mow your lawn, stop and take a long cool

think about exactly what you are doing. If you decided to give up the habit, it would not be long before the whole character of your precious greensward changed. It would not be only the daisies and dandelions that would run riot, but a whole gamut of plants from the neighborhood would invade the lawn as succession got under way. So every time you unearth the mower, you are in fact embarking on a controlled ecological experiment, the sole aim of which is to hold back succession. As succession is part of the process of evolution, you are, at least in one sense, holding back evolution—an act that you should perhaps not embark on too lightheartedly.

I quite enjoy mowing the lawn. The bit I don't like is having to pick up all the clippings—and that, after all, is the crux of the process of keeping a lawn well in hand. Without the intervention of the mower and the rake, part of the annual production of grass would fall to the ground to be stored in the form of humus, changing the structure and other properties of the soil. As soon as a perennial woody plant starts to grow within the community, part of each year's energy is stored in the form of the standing crop, which will protect and shade the surface of the soil. It is clear that succession is brought about, at least in part, by the storage of energy in the form of humus and standing crop, and that these together gradually change the habitat. It is also clear that you, plus your lawn mower, plus your compost heap, put a stop to all this in order to have a lawn. As long as you remain an integral part of the

97

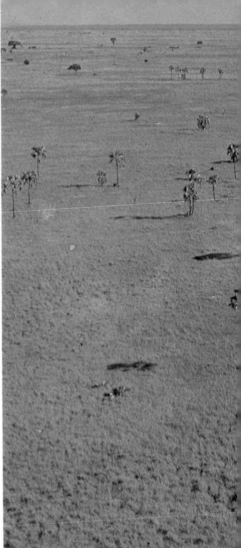

Whenever you mow your lawn you are in effect holding back the natural process of succession. The mower leads to the death of all those plants that cannot tolerate having their tops chopped off at regular intervals, leaving mainly grass.

living system and perform your gardening function each year, the lawn will remain a lawn—and thus, together, you and the lawn could be regarded as a climax ecosystem. In order to make it perfect, you would merely have to get back the energy you expend in mowing the lawn and earning the money to buy your mower by eating the grass clippings.

Now that you understand the full implication of your lawn-mowing, take a close-up look at the hallowed sward; it will surprise you just how many plants go to make up the community: *Nostoc*, mosses, puffballs, and two or three sorts of grass, together with a number of broad-leaved plants that consist of flat rosettes of leaves

appressed to the ground, with a central bud that is just waiting for you to go away and leave it alone so that it can get on with the business of growing and flowering. These broad-leaved rosettes include such weeds as daisies, dandelions, and shepherd's purse, and although their life form is helped by the regular pressure of the mower and the roller, this is the normal way they grow; a rosette of leaves, with a central bud that is held close to the ground and encloses the embryonic flowering spike. A lawn is therefore a very special type of ecosystem, in which the dominant plants are all plainly modified with respect to mowing pressure.

We normally think of lawns as specialized climax systems of the urban environment but they do have natural counterparts, which are called grassland or *savanna*. In nature, the active agents that remove the excess of the annual production and keep invading species at bay

are herbivorous animals and/or fire. There are a number of areas in the world where there is too much rainfall for it to be desert and too little rainfall to support lush forest and it is in such *tension zones* that grazing animals and fire can go to work. The hot, dry conditions of the desert are ideal for fire, but there is nothing to burn; conversely, there is plenty to burn in the lush forest, but it is too wet, and most of the kindling is well above ground level. Thus it is often in the areas in between that fire becomes an important factor.

One such area, which cuts across West Africa and includes half of Ghana and Nigeria and the north of Sierra Leone, is covered with grass and must be counted as one of the largest lawns in the tropics. The rainfall in this savanna belt is less than 40 inches a year. The adjacent zone of tropical rain forest may get only as little as 45 inches in the same time, so the difference does not seem much in absolute terms, but it is evidently

Grasslands may be thought of as enormous uncultivated lawns, which are kept down by fire and/or the grazing of animals. Although the factors that prevent the growth of trees vary, the result is always something like this tropical African savanna.

enough to tip the balance in at least one vital aspect: the effects of fire are small in the forest but very great in the savanna.

All the evidence suggests that man evolved on the African continent and spread out from there. There is also ample evidence that man has, to put it kindly, had the control of fire for more than 300,000 years. In West Africa, there is sound evidence of regular burning from about 25,000 years ago; however, some of these fires may well have been started by lightning. To support this theory, there is fossil evidence of extensive fires back in the Mesozoic period, at least 100 million years before the arrival of man. It is therefore difficult to come to a firm conclusion as to whether

In only a few days after a grassland fire has ravaged it the rains can revitalize the scorched land. The fresh green growth is dotted with brightly colored flowers such as this savanna lily.

themselves monsters by temperate-zone standards: they often consist of great tussocks between 3 and 10 feet tall.

Fire-tolerant shrubs and trees have very thick bark that allows the delicate inner living tissues to escape roasting, and in addition they are capable of reproducing by means of subsurface suckers. It is probably this reproductive ability that is of the greatest importance to their survival. To understand why, you need to have seen a grass fire. If such a fire is running before the wind, it is possible for a frightened animal that panics and runs back through the fire to emerge unscathed. The fact is that a grass fire happens very quickly. The line of fire is narrow and moves rapidly across the land. Because there is nothing bulky to consume, the fire doesn't burn for very long at any one point, and so it does not cause massive damage. The intensity of the fire is, therefore, a key factor and the exact effect of a savanna fire will depend not only upon the time of year when it occurs, but even upon the time of day. A fire early in the morning, when the dew is still on the ground and humidity is high, will cause as little destruction as one early in the season at the end of the rains. Later in the season and/or in the day, when everything has become tinder dry, things can be considerably worse.

Imagine that the annual dry-season fire has just passed. Everything is black and charred; there are no leaves on the trees, only plumes of ash swirling over hundreds of square miles of desolation. Yet, a month later—especially if it has rained in the interim—everything is green once more, with grasses and trees in full leaf and strong new shoots pushing up from the subsurface suckers. Soon, as a result of its remarkable ability to reproduce after being burned, the vegetation is bright with color, as a host of bulbs and other underground storage organs spring into life and throw up great heads of flowers.

Of all the plants of the savanna the grasses are the most important. Today, in fact, one might say that this is true on a world scale, because wherever man has destroyed forest, grass usually comes in and covers the destruction with greenery. The grass family (the Gramineae) is remarkable in that the plants are all capable of rapid regeneration after being grazed. The key to their success is that the bulk of the grass plant consists of leaves, and the leaves put on new growth at the bottom, not at the top like the stems of other plants; so they can withstand both chopping and

the West African savannas are natural or man-induced. They are certainly man-maintained.

It would be wrong, by the way, not to point out that most of these savannas are far from pure grassland. The range is from pure grass, which is a relatively rare phenomenon, through scrub savanna, which includes many fire-resistant scrub species that grow to a height of some 10 feet, to savanna woodland, which has trees 50 feet in height. The grasses that cover the ground are

chomping. Also, the flowers (most of which can hardly be called flowers) are wind pollinated, and—as every hay-fever victim knows—the wind spreads pollen around very efficiently.

While we are on the subject of getting around efficiently, this is, of course, one reason why the Africans burn grass. Elephant grass (so named for its size) impedes rapid transport, whereas walking through burned savanna, although messy, is much easier. Burning is also carried out for other reasons: to induce young, succulent growth for feeding cattle; as a first step in clearance for agriculture; and, of course, just "for the hell of it." It is well worth noting that any area of grass savanna that is protected from fire gradually reverts to woodland savanna, and even non-fire-tolerant trees start to come back. There is no getting away from the facts that fire holds this type of vegetation in its usual spectacular equilibrium, and that, as we have said, most fires today are started by man.

Whether man-made or natural, the savannas of West Africa are small as compared to two other great areas of natural grassland, the *pampas* of South America and the *steppes* of Eurasia. It is interesting that both words mean "plains" (one in Spanish, one in Russian). They refer to the flatness of the landscape, for both areas are gigantic plains. The steppes, which stretch from Hungary almost to China, must rank as the world's number one lawn. One of the factors that combine to preclude the growth of trees in such grasslands may well be the characteristic high winds, related to the absolutely flat terrain. In the steppes, the winds, combined with the long, cold winters, the short, hot, arid summers, and the unstructured soils, make the natural development of forest impossible over vast tracts of land. And something similar is true for the pampas (although in their case, it has been argued that pre-Columbian Indians may have destroyed the original forest).

As you can see, we do not know the exact reason or reasons for the existence of savannas. But wherever natural grasslands do exist, we can certainly regard them as experiments in living that evolution has not been able to take very far—at least, not yet.

The summer-dry scrub forests of the Mediterranean are, to the tourist or holiday maker, relatively unknown. What do *you* think of when you hear or read the word "Mediterranean"?

If you want to understand why the inhabitants of West African savanna regularly set fire to the elephant grass that grows there, just imagine what it must be like to walk through it.

Idyllic holidays beside a clear blue sea? The cradle of at least two civilizations? Whatever it is, I should like to bet that vegetation doesn't get a mention. If you ask the average person who has just come back from, say, Italy or Greece, to comment on the vegetation, he will more often than not respond with a blank look—followed perhaps with the words "vineyard" and "olive groves" thrown in as an afterthought. The fact is that most people go to the Mediterranean during the hot, dry summer, when typically the region's vegetation is at its brownest. This is a land of about 24 inches of rain annually, with a marked period of summer drought; a land of summer hibernation—or *aestivation*, as it is properly called. So one might expect summer deciduous forest to dominate the scene. Yet the dominant forests are evergreen.

The region has not come into popularity with *Homo sapiens* only recently, of course; it has in fact borne the brunt of man's endeavors over a very long period of time, and just about every cultivable square yard has been cultivated. Any tracts of land that are too steep or stony for agriculture support rough scrub, which is grazed to the ground by a full cross section of livestock—but especially by sheep and goats. When it comes to destroying vegetation, the latter are the most damaging, because goats will eat almost anything in the way of plant material, whereas sheep are much more fastidious nibblers and pick at the vegetation in a more selective manner. The end result is much the same, however, for the sheep and cattle graze out the more tender plants to a point where, in time, the vegetation is useless for everything *except* goats.

The plants that make up the surviving vegetation of this region have two special adaptations that have ensured their survival. First, they are xeromorphic. Secondly, many of them are armed with an astonishing array of prickles and spines to ward off the attentions of would-be grazers.

The best time of year to see the flora is in the spring, when the color is astounding. A springtime visit to one of the famous temples on the Greek islands, which become hot dust bowls in summer, is a never-to-be-forgotten experience. (I have a sneaking suspicion that some of the ruins are not ruins at all, but were placed there by some imaginative landscape gardener in order to "set off" the flowers.) Two types of plant dominate the spring flora: *geophytes*—that is, plants that lie safely hidden in the soil for most of the year as bulbs or corms—and *annuals*, which accomplish their period of aestivation as seeds. Both types have showy flowers.

In the summertime, however, what you see is a far different picture: dry, parched scrublands and

Below: these field poppies germinate, flower, and fruit quickly to take advantage of the short period of the Mediterranean spring, before the arrival of the long, hot, dry summer. Right: a line of gnarled old olive trees in a grove on the Greek island of Corfu. Nobody who has visited the Mediterranean lands can fail to recognize the dusty green, drought-resistant foliage of such trees.

summer-scorched fields, with only three types of woodland coloring occasional parts of the landscape green and producing cool, welcome shade. First of these small forests are the olive groves, which are entirely man-made and very widespread; although not everyone would call them handsome trees, they are certainly of key importance (in fact, in certain parts of the region, a person's wealth is judged by the size of his olive groves). Secondly, there are fairly extensive pinewoods, which shade a rich flora of tree heather, lavender, and small evergreen oaks. Wherever the tree layer is destroyed, the undershrubs can take over, and the resultant vegetation is known as *maquis* or *garrigue*. The extent of this scrub can be gauged by the fact that the French resistance fighters in World War II took their name from the vegetation because it afforded them such ample protection. Very prominent in the scrub vegetation are the small-leafed, evergreen oaks that were mentioned above. The most typical of these is the holly oak, whose serrated leaves can make progress through the scrub a painful experience. Here and there, however, larger oak trees emerge from the scrub canopy and there are enough of them in places to form the third type of woodland.

These larger oaks are readily identifiable by their very thick bark, which, more often than not, will have been stripped off from the trunk, making it look not unlike a well-groomed French poodle. Human beings have done the stripping, for these large trees are cork oaks—the stock-in-trade of the cork industry—and their bark is the stuff that makes champagne possible. In nature, a thick corky bark makes a tree more or less immune from the effects of fire. Thus, the presence of cork oaks indicates that fire is a hazard, if not a main factor, that determines this type of vegetation.

Cork is remarkable stuff. I am speaking now not specifically of bottle stoppers, but of the cork that comes in all shapes and sizes, from the thin, paper-like variety found on birch trees, through the deeply fissured types such as on the olive, to the "champagne" of corks, *Quercus suber*, the scientific name for the cork oak. Cork is defined as dead tissue produced to protect the living tissue of a tree trunk from damage, especially from disease. The starting point of all tree trunks is a young, green twig. As it increases in length, it

This cork oak has been stripped of most of its thick covering of bark, which serves to protect it against both water loss and fire.

must also increase in girth, in order to keep its strength/weight relationship within compatible limits. The young twig is protected by the layer of cells called the epidermis, which is perforated by stomata. As the twig thickens, the epidermis is put under increasing strain, and it would eventually rupture and lose its protective function—which could mean death for the plant—if it were not for the formation of bark. Before the protective "skin" is broken, living cells below the epidermis take on a new function and begin to divide, producing more cells to keep pace with the increase in girth. The cells produced in this way go to make up the bark; and successive layers of bark are formed as the tree continues to grow. Underneath these new layers, the old tissues die; the thick dead tissue making up the greater part of the bark is what we call the cork.

As anyone who has ever looked at trees knows, there are many types of bark. The paper (or silver) birch, for instance, produces a complete new cylinder of bark each year; this thin sheath is known as *ring* bark. In contrast, the fissured barks are produced from overlapping plates of dividing cells. Although watertight, bark is supplied with numerous pores, at least some of which correspond to the original positions of the stomata in the epidermis. These pores are called *lenticels* and traverse through the cork mass, even though it may be eight inches or more in thickness. It is these pores that allow the cork to be compressed, and thus make it an efficient air-tight stopper for corking bottles.

The cork oak has especially structured leaves, which save it from desiccation during the hot, dry Mediterranean summer. The broad leaves—especially the underside, on which most of the stomata are located—are covered with a dense weft of hairs that trap a layer of moist air close against the surface of the leaf. Water vapor passing out through the stomata is, to put it simply, impeded by an effective barrier of air that is already laden with vapor.

It is somewhat debatable whether Mediterranean scrub-forest vegetation should be considered as one of the main vegetation zones or as a special type. Certainly this kind of vegetation is not restricted to the region of the Mediterranean Sea. In the south-western states of the United States, for example, a similar vegetation type—sagebrush, or chaparral—is very extensive. It is an interesting fact that in the 17th century the chaparral supported a very large population of Indians, averaging about eight per square mile. They lived by hunting, but a staple factor in their diet was acorns. The natural population of herbivorous animals, which might have denied the food supply to the Indians, was held in check by grizzly bears, which roamed the sagebrush. As time went on, however, man gradually eradicated the grizzly, the herbivores increased in number, and the vegetation was destroyed.

It would thus appear that the main *natural* limitation of such areas is summer drought, and the climax vegetation is probably some form of xeromorphic forest. What is difficult to ascertain is what the vegetation would be like without the intervention of man and—especially in the Mediterranean regions—his goats. In certain areas around the Mediterranean Sea, where 20th-century tourism has brought increased employment to the region and the peasants have left their pastoral duties to work in hotels, the broad-leaved forest seems to be coming back into its own. And so in time we may learn exactly what the potential natural vegetation of these summer-dry lands really is.

Twenty-four inches of rain, properly spread throughout the year, is enough to make the Mediterranean bloom; less than that spells doom for almost any plant that attempts to live there.

Note the important word "almost." One definition of a desert is that it is an area devoid of life; yet all of us know that cacti grow in deserts—which makes something of a mockery of the definition. Still, definitions matter a great deal to botanists. Let us digress from our main subject for a moment to explain why. When a botanist has found a brand-new sort of plant, one that has never been described before, he wants his fellow scientists to know about it. So he must define it for them: he describes it in a formal way, gives it a Latin name, publishes the description and lodges original or "type" specimens in one of the leading *herbaria* of the world. Herbaria are places where collections of dried and preserved plants are kept, and where bands of dedicated *taxonomists* (scientific classifiers) work on the classification of the plant kingdom.

In some ways herbaria are akin to patent offices. They are patent offices of plant evolution, which

The giant saguaro cacti of Arizona. The enormous green stems, which can be up to 50 feet tall, are living water storage tanks.

106

not only file the "patent" and the "invention" but carry out further developmental research. In addition, they produce *Floras*—which in this context are not plants themselves but catalogs of the plants that grow in specific areas. A good Flora is a handbook of evolution, not only of the plants themselves but also of ecosystems, because it would be an impossible task to describe vegetation without a basic knowledge of the different types of plants involved. Thus, the work of a herbarium is immensely valuable to all students of botany and ecology. It is tempting to coin an aphorism: To be able to classify is to begin to understand. And that is why definitions matter. Now let us try to define—in other words, to understand—desert vegetation.

Because of their fleshy stems the cacti are difficult types of flowering plant to store in herbaria. And among the cacti that liven up the deserts of the New World, the giant saguaro is the most difficult to store because it is the largest. This prickly monster can grow to a height of 50 feet or more and can weigh as much as 10 tons, although it may take as long as 200 years to reach such dimensions. Cacti (which are called *succulents* because of their watery tissues) can be summed up as living water storage tanks, each joined to an extensive rooting system that spreads out through the surface layers of the soil ready to pick up the rain as soon as it comes along. Immediately after the rain, the internal tissues (which are modified to store water) are full, and the cactus is plump and turgid; it then gradually uses up the water, becoming increasingly limp and flaccid, until the next rainy day. In a good year—that is, one in which there is a lot of rain compared to the norm—the "barrel" can grow quite rapidly. But this is of rare occurrence in most deserts, and so life is mainly at cactus pace.

The only visible modifications for life in the dry climate are a thick covering of wax and a complete absence of leaves. In place of leaves, cacti have spines borne on special organs called *areoles*, which, sometimes together with bristles, hair, or wool, make them look like furry pincushions. The areoles are a special feature of the cactus family. The spines help to ward off would-be herbivores, which could remove not only parts of the precious annual plant production, but also

Left: following an abundant rainfall, the "barren" desert soil springs to life. Here sand verbena and evening primroses transform a bleak landscape near the Colorado River in California.

Above: the "crown of thorns" may look like a cactus, but the peculiar red flowering head and the sticky white latex that bleeds from a broken stem identify it as one of the spurges.

the protective waxy cuticle, and this would lead to excessive loss of water. In the absence of leaves, the whole shoot takes over the function of photosynthesis; that is why the great, fluted stems are green with chlorophyll. If you examine the depths of the furrows carefully with a lens, you will find the stomata. But do be careful because the spines are very efficient weapons.

The main weapon that the cactus has against the environment is the ability of its stomata to work the night shift. In the often bitter coldness of the desert night, a cactus opens its stomata,

109

allowing carbon dioxide to diffuse in with minimal loss of water vapor. Because there is no sunlight at this time, the carbon dioxide cannot be immediately turned into sugar, and so it is stored in the cells as an organic acid. As the night progresses, the tissue of the cactus becomes more and more acid. In the morning, when the sun rises, the stomata close, thus preventing massive loss of water in the heat of the day. The organic acid is used as a source of carbon for sugar production, and the cactus gradually cures itself of its self-inflicted acid indigestion.

This biochemical survival mechanism is called *crassulacean acid metabolism*. It gets its name from the fact that another plant family, the Crassulaceae (or stonecrops), which are also succulent plants with water-storing tissues, also work the stomatal night shift in dry places. Faced with the same problem, both families have solved it in the same way. It is among the desert succulents that we find the most clear-cut cases of *convergent evolution*—that is, evolution of similar shape, structure, and even biochemistry in unrelated families, in order to achieve survival in a particular type of extreme environment. A perfect example of this is that of the Cactaceae (cacti) and the Euphorbiaceae (or spurge family): it takes careful examination to tell certain members of these two unrelated familes apart; from the outside they look almost identical. The easiest way to distinguish between the two is to break off part of the plant, when white latex will flow freely only if it is a spurge. Remembering the spines, however, perhaps the best thing to do is to wait until the plants flower, for the two families have very different flowers. Cactus flowers are marvelous things and well worth waiting for, but unfortunately they do not last long. They are pollinated by insects (some by bats) and then produce handsome fruits, some of which are very edible once the prickles are removed.

Cacti are plants of the New World. Wherever you find them on the other side of the ocean, they have been introduced by man. The great, straggling, aptly named prickly pear makes just as effective hedges in Italy as it does in its native home of Central and North America. Not all cacti live in deserts—some even inhabit tropical forests, growing as epiphytes on the trees—and not all deserts are the romantic sort of rolling dunes that you sometimes see in movies. Most of the world's deserts are stony, inhospitable places: they are not ideal year-round situations for an energetic botanist. However, wherever there is a stony surface, run-off of the meager rains can supply the valleys with sufficient water for a dry-scrub forest type of vegetation.

One rather surprising thing about the vegetation communities of these semi-desert valleys is that the plants are often very regularly spaced. In places they can look as though they have been planted by a geometrically minded gardener. The explanation could well lie in the search for water, with each plant drawing on the resources of a certain area in order to obtain the amount it needs. There is, however, another possibility: it may be that the plants produce toxic substances that prevent seed germination and the growth

Right: the reddish seed pods of this Mexican spurge show up very clearly against the gray-green of its succulent stems.

Above: when conditions are just right, cacti produce flowers that are well worth waiting for. The showy blooms attract insects, which help to bring about the essential process of pollination.

Left: the ribbed green stems of snake cacti are well furnished with chlorophyll and stomata, for they are the organs of photosynthesis. The leaves are reduced to spines, which protect the plant.

111

of seedlings in their vicinity. If such a mechanism of competition (called *allelopathy*) does exist, it could account for the regular spacing.

Whether there are or are not such toxins, the desert soils do, in places, hold many secrets about plant survival. The most spectacular of these is the soil's ability to retain hoards of the seeds of many types of annual plant—seeds that are resistant to both decay and desiccation, and that, after remaining hidden for years, can germinate following rain and carpet the barren soil with flowers. These plants are called *desert ephemerals*, and they can crowd their whole life cycle into the few short days of an occasional desert spring. One fascinating thing about them is that all the different sorts do not come up after each rain; some are dependable performers, but others seem to require special conditions to stimulate their emergence. This can make plant hunting in the desert a very exciting process, because you never know exactly what you may find. And, when you have found it, it may be a case of back to the Latin homework for the discoverer.

We turn now to saline deserts. We have discussed deserts without water and (Chapter 5) ice deserts made of solid water. To round off the problems of evolution under water stress, we may refer to deserts that are saturated with water, but that remain deserts because the water contains too much salt and is thus unsuitable for most plants.

A living cell consists of a protoplast of structured biochemicals surrounded by a double membrane that has some remarkable properties, not the least of which is its *semipermeability*. This means that the membrane regulates the passage of molecules through itself, the regulation being based—at least in part—on molecular size. Small molecules, such as water molecules, can pass in either direction with relative ease, but the membrane presents more of a barrier to the passage of such larger molecules as sugar. Modern research indicates that the actual function of the cell membrane is much more complex than this, but for our purpose, let us stick with the single quality of semipermeability. A living cell is thus a special sort of bag, which is full of a dilute solution of chemicals (including a minute amount of common salt) and a mixture of complex biochemicals. If it is placed in a strong solution of common salt, both salt and water molecules will be exchanged through the membrane in order to redress the difference in concentration that exists on either side of the membrane. Because water can pass the barrier more easily than the salt, water moves out of the cell at first faster than the salt moves in; the cell therefore shrinks. In time, though, the salt molecules do get through the membrane and bring the system back into equilibrium as water passes back in; and so the cell once again becomes turgid. At equilibrium the concentration of salt inside and outside the cell should be the same.

Whether or not a true equilibrium is reached depends on other properties of the membrane, which do not concern us here. But one important special feature is that the living cell is capable of taking up certain minerals against a concentration gradient. This is indeed remarkable, because one of the basic laws of the universe states that all systems tend toward an equilibrium; therefore, energy must be expended if a state of disequilibrium is to be maintained. Thus a living cell must expend energy in taking up minerals that are in very low concentration outside—such nutrients as phosphate and potassium—and maintaining them at a much higher concentration within their tissues. And, conversely, any plant that is growing in a situation where there is a vast *excess* of one mineral, such as common salt, must expend a lot of energy in keeping it out of its tissues, or in getting rid of it once it has accumulated. The whole operation revolves around the related problems of water supply for the living tissues and the process of evaporation.

When water evaporates it changes from a liquid to a gas, and a lot of energy is used in bringing about the transformation. This is one reason why we hot-blooded animals sweat. Sweat is a dilute solution of salt and part of the energy used to evaporate the sweat comes from our skin, thus cooling our heated brows. The trouble is that only the water goes; the salt is left behind—which is why the human skin tastes salty. Now put yourself in the place of a plant, with your roots embedded in a saturated saline soil and your shoots blowing in the hot, dry wind. Water, evaporating from the living cells, passes out by diffusion through the stomata, and more water molecules move up in an endless cohesive chain to take their place. If even a dilute solution of salt comes in at the bottom, it is going to be concentrated by evaporation at the top. As the bulk of the evaporation cannot take place on the surface of the plant, which is covered with a watertight cuticle, it must take place from the living cells, inside the

Above: the cord grass, Spartina, can rapidly transform salt marsh into dry land. It is one of the very few plants that have evolved to spend their lives in the extreme conditions of a coastal marsh that is flooded by sea water at each high tide.

Right: like the dry-desert cactus, this salt-desert Salicornia is a succulent plant, with reduced leaves, a thick, waxy cuticle, and juicy, water-filled tissues. Most salt-desert plants resemble dry-desert plants, too, in that they can be grown in "ordinary" soil as well as in the special habitat within which they have evolved. But salicornias cannot be raised in a botanical garden or anywhere outside their natural harsh environment

Eelgrass, one of the aquatic turtle grasses, is really a pondweed, and grows and flowers submerged in shallow salt water.

leaf, that line internal air spaces that join up with the stomata. This means that any concentration of salt will occur within the living cells themselves—an acute problem, which the plant can solve only by expending energy, either to keep the salts out of its tissues at the bottom or to get rid of them at the top.

Evolution appears to have resolved the problem in saline deserts by producing the *halophytes*, or salt plants, which share many of the modifications of the dry-desert succulents, including reduced leaves, a thick waxy cuticle, and juicy, water-filled tissues. All these modifications would appear to reduce the amount of water that flows through the plant in its lifetime and, hence, the amount of energy that must be expended in dealing with the problem of too much salt. It seems such a nice explanation of the construction of the halophytes—but botanists have a hard time trying to get data to prove it.

It is interesting to speculate on the evolution of this mechanism for dealing with too much salt. Algae that live in the sea are never exposed to the problems of evaporation because they live submerged; and those that live between the tides have a thick, sticky outer covering that minimizes water loss during the time when the tide is out. Highly saline conditions are not confined to the coast, however. Large areas of arid desert have saline soils as a result of being almost continually subject to evaporation, which concentrates dissolved salts in the surface layers. It may well be that the halophytes evolved under dry-desert conditions inland, where some are still found, then returned to and found a suitable habitat in the maritime salt marshes, where they became specialized xeromorphs with their feet in water. The fact is that because the seaweeds did not evolve roots, they could not make a real success of life on the shifting silts between the tides. So such areas were an open habitat—free for colonization, as it were—the properties of which could be exploited by any plant that had both a root system and the attributes needed to overcome the salt problem. It was in this way that the halophytes must have arrived to liven up the coastal scene.

Very few of the halophytes have evolved complete dependence on a saline habitat. The notable exceptions are a group of plants called *Salicornia*, or glassworts, which are very specialized members of one large plant family, the Chenopodiaceae. Although we don't understand why, the glassworts are so dependent on their saline habitat that they cannot grow in ordinary soil. Here is an interesting description of this family, quoted from a modern Flora: "Annual or perennial herbs or shrubs, rarely tree-like, frequently succulent, or with bladder-like hairs which give the plant a mealy appearance. About 75 genera and at least 500 species. Cosmopolitan, but mainly in arid regions." (Families of plants are divided into groups called *genera*—the plural form of *genus*— and each genus is divided into species. The basis of the division is the similarity that the plants bear to one another.) The Flora makes the following statement about the genus *Salicornia*: "Probably about 50 species, commonplace in saline districts. Many of the species are critical [which means they are difficult to tell apart] and need much further work for their elucidation. Herbarium specimens, dried in the usual way, are almost useless." So any botanist who wants to become famous for his work on the glassworts will have to work hard at his taxonomy, preferably somewhere on the seacoast, because fresh specimens are required and, as we have already said,

they cannot be grown in normal garden soils.

Other halophytes will grow quite happily in garden soil; in fact, many of them do much better there than in their usual haunts. The only precaution you must take if you want to raise them successfully is to keep your plot well weeded, because most of them cannot withstand competition, especially from ordinary plants. Again it is tempting to ask why and to put forward a hypothesis. A simple explanation would be that plants that are modified to reduce the uptake and loss of water will also grow at a slower pace, and will therefore be at a disadvantage when growing among unmodified plants. A plausible theory, but the problem is to prove it!

One group of flowering plants have gone even further than the glassworts: instead of just paddling about on the edge of the sea, facing all the problems of water, salt, and evaporation, the

A young living Arctic landscape. The many pools are being engulfed by peat and colonized by a succession of green plants.

Peat-cutting in Ireland. Peat is a natural source of energy—the excess energy of many thousands of years of photosynthesis.

turtle grasses have taken the plunge and become subaquatic. The turtle grasses form extensive submarine meadows and grow between the tides in estuaries and salt marshes and on coral reefs, binding the mobile silts and sands together with their long roots. The living tissues of the submerged plants must come into equilibrium with the concentration of salts in the seawater. However, they are never exposed to the problems of water loss and hence to further concentration of salts by evaporation. For this reason their leaves need little or nothing in the way of a protective cuticle, and without a cuticle they are able to take up carbon dioxide from the water over their whole surface areas. The turtle grasses are thus perfectly adapted to the aquatic way of life.

The turtle grasses not only grow but also flower under water and it is only when they flower that it becomes obvious that they are not really grasses at all. In reality they are members of special families that are closely related, not to the grass family, but to the pondweed family (Potamogetonaceae). The flowers produce very special elongated pollen grains, which water currents carry to female flowers. In addition, the turtle grasses have the great advantage that they are easily propagated by pieces of the plant that break off and float away. This quality is ideal for a plant that lives in shallow sea water, where it is subject to wave action. To gauge the success of the turtle grasses, do a bit of beachcombing along the shores of the Mediterranean or the Florida coast just after a storm, and you will find enormous rafts of their leaves cast ashore.

The turtle grasses (which are, incidentally, browsed on by turtles) are found throughout all the major regional climates. But they really come into their own in waters, especially in the tropics, where they are important components of coral-reef ecosystems. There is, though, one peculiar, and as yet unexplained, gap in their distribution: for some reason they are almost entirely absent

116

from the coastal waters around South America.

We have seen that too little water presents a problem for living things. So does too much water, even if it is of the right sort, that is, not too salty. The problem is that water and oxygen do not mix very well. Oxygen, in fact, is only slightly soluble in water; and once it *has* dissolved, it diffuses very slowly. A layer of water, especially if stagnant, will thus present a barrier to oxygen. And because most living organisms require both water and oxygen—well, thereby hangs the tale of the formation of the extensive northern peatlands.

Wherever rain falls on the surface of the earth, it possesses a certain amount of potential energy, which is lost as it flows down toward the sea. The loss of this potential energy causes the erosion and dissolution of the rocks that make up the catchment area for the water. It is in this way that the almost pure rainwater picks up an increasing load of dissolved minerals, including all the nutrients required for healthy plant growth. While it is flowing rapidly, the water is also charged with oxygen, which is continually being "stirred in," thus producing an ideal culture solution for any plant that has the adaptations necessary to enable it to hang on.

In places where the rate of flow is reduced to such a level that the water body can no longer carry particles along in suspension, plant life can gain a firm foothold, organic matter will accumulate, and oxygen-deficient conditions will begin to prevail as the process of succession gets under way. The organic matter laid down under these special circumstances—a great deal of water and too little oxygen—is called *peat*; and the plant communities that not only develop on, but form, the peat are called *mires*. Wherever the right conditions prevail on earth, mires or similar ecosystems await discovery.

The most exciting journey of discovery I have ever made was from the most northerly part of Canada on Ellesemere Island in the Arctic, to the south of the Great Lakes. It was made, as are all the best journeys, with a companion—my friend Bill Radforth, a botanist who, like myself, has spent much of his life discovering mires. (Being a Canadian, he calls them *muskegs*, but that is the only difference.) The exciting thing about this particular journey was that it took us not only across 3000 miles of terrain, but through about 10,000 years of history. As we traveled southward the landscapes told the story of a gigantic experi-

The world-champion peat formers are the bog mosses. Two of the many different varieties are shown in the picture above.

ment that has been going on ever since the ice sheets of the last ice age began to melt.

Our journey started on the Arctic ice cap, which, apart from a few migrant visitors like ourselves, was devoid of visible life. In certain places, where fresh snow was melting, we found the unmistakable pink of "red snow." The ice caps that still cover the higher parts of the Arctic islands had for a long time been melting, a fact that is manifested by the rapid rise of the landmasses as the great burden of ice is lifted from the earth's crust. Once this burden has gone from a given place, the bared, ice-worked landscape is a dry desert; for a few weeks each year, however, it flows with meltwater and becomes a wet desert.

Just as the annual cycle of freeze and thaw molds the life of the snow alga, so it molds the face of the landscape, producing the geometric patterns that typify the high tundras. The first type of vegetation to fill in the depressions between these ice-worked patterns is embryonic mires. Once these have become established, the annual deposits of peat retain more and more water, which allows them to spread. And as they grow they obliterate the terrain features. Riding south from the edge of the Arctic ice sheet, you find yourself in fact riding two time scales. The first is the time period that has passed since the landscape was entirely covered with ice. The second is related to regional climates. The farther south you go, the longer is the annual period of sufficient warmth for plant growth. The two time scales are interrelated—so much so that their

effect can be summarized by the statement that the potential for the development of living systems grows greater as you go farther south.

Thus, the landscape patterns change from the skeletal forms of the embryo mires, through which the bare bones of the ice-worked landscape protrude, to something that Bill Radforth has christened *terrazoid*. From the air, the terrain is seen to consist of plateaus of peat, separated by water tracks, ponds, and lakes. As the peat plateaus grow and expand, they contain the rain-charged meltwaters that are flowing through the terrain to the sea. The peat mass acts both as a dam, ponding back the water as it flows through

Below left: in the permafrost region south of the Arctic ice sheet the Canadian terrain is covered with peat-forming vegetation, which from the air looks like a vast sheet of polished marble. Below center: toward the southern limit of this permafrost zone, the warmer climate enables evergreen trees to colonize the edges of the great spreading plateaus. Below right: south of the permafrost insufficient rainfall in the warmer, longer summers confines the formation of peat to depressions between the watersheds that are supplied with groundwater throughout the year; the higher areas between the depressions are dry and therefore forested.

the land, and as a reservoir, holding a certain amount of water within its substance. This is especially true on the plateaus, which soon become dominated by the king of the peat organisms: the bog mosses (*Sphagnum*).

Although, as in all mosses, the leaves of *Sphagnum* are a single cell thick, they consist of two types of cell—narrow, living cells containing chlorophyll, and much bigger, empty, dead ones. The dead cells are well supplied with large pores, which connect the interior of the cell to the outside world. The big cells thus act as internal reservoirs; and, as each moss plant consists of masses of leaves, the whole thing acts like a sponge. It is in this way that the bog mosses can grow into their characteristic hummocks, thereby raising both themselves and the process of peat formation above the groundwater.

Once the bog mosses have established their watery presence, the process of mire landscaping goes on apace. The peat plateaus become lobed and grow out to fill and obliterate the smaller ponds and lakes. From high up in the air, the lobed masses look not unlike great sheets of polished marble, and the term *marbloid* is used to describe them. The illusion of marble is heightened by the fact that predominantly white fruticose lichens grow among the sphagnums, giving the vegetation a mottled appearance.

The marbloid muskegs of Canada cover vast tracts of land. One could probably walk clear across Canada and, apart from swimming the odd river, never step off peat. (It would be a long and devious route and I don't know of anyone who has tried it, but I must confess that it is one of my secret ambitions to have a go at it.) The immense size of the peatlands is surprising when it is borne in mind that much of central and northern Canada gets less than 12 inches of annual precipitation. It should be a cold, arid

desert; yet it is covered with peat—a type of vegetation that is absolutely dependent on enough water to keep it saturated. The explanation lies, in part, in the evolved relationships between the vegetation, the climate, and the permafrost.

From the air, the lichen-rich marbloid muskeg looks as if it were partly covered with a dusting of snow, and the reflection from this white surface in the summertime can be very painful to the eyes. The presence of the lichens thus increases the reflectivity of the terrain and consequently reduces the amount of heat absorbed by the peat. As a result, the permafrost melts only very slowly, and the active layer remains shallow; moreover—and this is of key importance—it remains charged with meltwater, thus allowing the process of peat formation to continue.

At the southern edge of the belt of marbloid muskeg, the landscape pattern changes once again, for there is now sufficient potential for the growth of trees, and so the terrain takes on a stippled appearance from the air. Here we arrive at the southern boundary of the great sheet of permafrost and, hence, at the end of the dominance of muskeg across North America. In the absence of permanent ground ice, there is no water-charged active layer to nurture the peat throughout the much longer summer periods. In those areas where there is little rainfall during the growth period, the continued development of peat is impossible. However, close to the coasts, in British Columbia, Newfoundland, and Nova Scotia, where the rainfall is very high, large areas of land are still covered with continuous tracts of uniform muskeg with little or no pattern. These are best called *blanket mires*, to distinguish them from the marbloid mires of the permafrost zone. Such mires are found throughout the temperate zone wherever the rainfall is in excess of 40 inches per annum. They are common features of the Atlantic seaboard of Europe, covering large areas of Norway, Scotland, and Ireland.

Farther inland, where the rainfall is insufficient to support true blanket mire, the development of peat is gradually confined to situations in which mires are supplied by flowing groundwater throughout the year, leaving the dry valley slopes covered with forest. These extensive groundwater-fed mire systems develop a peculiar surface

The world's best-known subtropical mire system: the Florida Everglades, where peatland, mangrove swamp, and estuarine marsh merge. It has a large plant and animal population.

structure, consisting of elongated hummocks or strings of bog moss, separated by elongated depressions and pools, all arranged at right angles to the slope of the surface of the mire. The strings act like a series of coffer-dams, impeding the flow of water through the system. From the air, these mires look like a network of strings and hollows. In central Finland, where they are dominant features of the landscape, they are called *aapamires*, which means "mires consisting of alternate raised strings and pools."

South of the string mire zone, mires become increasingly rare features of the landscape, because the long, hot summers of the subtropics affect the water balance of the catchment systems. Eventually, mires are found only in estuaries of large rivers, where they merge with coastal mangrove swamps.

The most famous example of such a subtropical mire system is the Everglades of Florida. Similar estuarine and coastal mires also exist in Cuba. Thus, peatlands span all the main regional climates of the Northern Hemisphere. The mires are, indeed, a gigantic experiment in the evolution of ecosystems that develop under the stress of too much water and too little oxygen. (A similar zonation of mire type exists in the Southern Hemisphere, although it is less well developed for the simple reason that there is little land at the right latitudes.)

With 370 million acres of this dry earth under a saturated blanket of peat, there can be no argument as to the success of the experiment. The world's reserves of peat are 295,000 million tons of organic matter, representing 1,700 million billion calories of stored energy, holding 39,600 million gallons of water. That is a fantastic natural source of energy, which man is burning up at an inordinate rate, mainly for generating electricity. No matter how the peat is used, the end point is the same: burning it oxidizes it into carbon dioxide, which is released into the atmosphere. If all the world's peat were oxidized in this way, it would produce some 447,000 million tons of carbon dioxide. So it is a fact worth thinking about that the world's mires are important indicators of the potential of this live earth.

The patterns of peat that unfold during the journey south from the polar ice cap are a unique record of the potential of the various landscapes to evolution, and to man. One can only hope that in our search for sources of energy, we shall not lightly set about destroying that record.

Man and Plants

Whether they like it or not, all human beings are herbivores, even if not exclusively so. I know plenty of people who are proud to be *absolute* herbivores, though they have a polite name for themselves: vegetarians. I have never yet met anyone who claims to be an absolute carnivore; and, in any case, such a claim would be an idle boast, because fish, fowl, and flesh are neither more nor less than reconstituted plant. Everything you eat, in other words, has vegetable connections. As the Bible puts it, "all flesh is grass," for ninety-nine-point-nine-plus per cent of the energy that flows through the living world is derived from the sun, and it is chlorophyll-containing plants that convert the bulk of this sun energy and store it in the form of starch for later use. The process by which they do this is known as photosynthesis. A small, yet important, part is fixed by *photosynthetic bacteria*, which, although they contain a special type of chlorophyll, live in and maintain an oxygen-free environment. They thus exploit the potential of environments that are closed to the major forms of photosynthetic plant life. The best place to see these unique organisms at work is in the coastal marshes and reef flats of the tropics, where they form pink crusts over the silts and rocks. If you break a piece of the crust, you will often smell something like rotten eggs, the trademark of oxygen deficiency, and the underside of the crust will have cherry-purple coloration. What you are looking at is the purple bacterium, *Rhodospirillium*, a bacterium that can enjoy the best of both worlds. It can live by photosynthesis alone, or, when growing in the dark, it can feed as a decomposer, on the energy in organic matter.

The minute amount of energy that is not derived from the sun comes in through another group of bacteria—the *chemosynthetic bacteria*. As their name suggests, these derive their energy from simple chemical reactions. If you keep your eyes open on a country walk, you will probably see at least the final result of the peculiar life

Because many animals are herbivorous, it seems only fair that a few plants should retaliate by being carnivorous. The leaves of the Venus's-flytrap close over small insects that touch their sensitive trigger hairs, and absorb the products of decomposition.

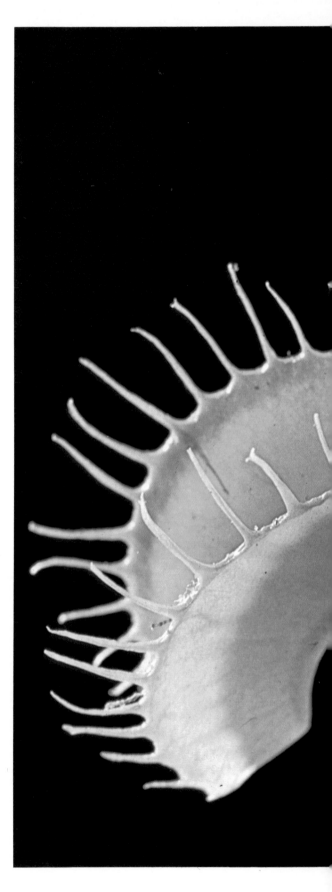

processes of chemosynthetics of one type, which obtain energy as a result of reactions involving iron. Visible evidence of their presence in ditches and woodland flushes are flocculent masses of what looks like rust-colored jelly. You are not far wrong if you think you are seeing rust, because you are looking at iron oxides, which have been precipitated out from solution as the bacteria have grown. However, as the bacteria are more often than not classified within the plant kingdom, the small contribution of chemosynthetic bacteria to the energy of the living world does not detract from the truth of the biblical statement that "all flesh is grass."

The only problem with being a carnivore is that, however efficient your method of catching prey, the laws of the universe state that you will be a much rarer type of organism than the herbivores you are eating. Energy can be neither created nor destroyed; it can only be changed from one form to another, and at each transformation there is a loss of potential energy as heat. Sunlight to plants, plants to herbivores, herbivores to carni-

Fill your plate with meat, fowl, and fish, and you are nevertheless still eating vegetable matter, because plants are the first link in every food chain. The photographs below show various plant-eating animals (from left to right): sheep, guinea fowl, and carp.

vores—there is a loss of potential at each step in the food chain. To live the life of a top carnivore—that is, an animal that eats only the flesh of other carnivores—simply moves your room at the top one step farther away from the energy source.

Man, whether a vegetarian or a carnivore, is thus ultimately dependent on plants for all his food requirements. And until the plastics age reared its polymerous head, the same was true for most of his raw materials. Even for plastics, in fact, plant life is essential. One of their main ingredients is derived from petroleum, and much of the energy required for their manufacture is derived from coal. Both petroleum and coal represent the excess of photosynthesis of the primary producers of long-extinct ecosystems. And so there is no getting away from the fact that plants are just as vital to the 20th century as to all past ages.

When it comes to staple diet, the world's population is entirely dependent on the members of one family of flowering plants: rice, corn, wheat, barley, oats, and millet, all of which are members of the grass family. Yet, dependent as we are on the productivity of this enormously important family, we do not make use of the bulk of the standing crop that is produced from the world's vast annual acreage of land under cereals.

We use the seeds, but even in modern high-yield varieties, the seeds represent less than 10 per cent of the total annual production. Some of the remainder is profitably fed to livestock, but the bulk is left to rot on the land. With two thirds of the human population living on the brink of malnutrition, this loss of stored-up energy should not be taken lightly (even granting that the value of the organic matter to the maintenance of soil fertility cannot be overestimated). It is particularly sad that, in general, human beings eat very few leaves, for leaves are full of nourishment.

If we reflect upon all the plant material we eat, we shall see the truth of this. We eat fruits, seeds, stems, roots—but what leaves are to be found in our diet? Spinach, cabbage, sprouts, and lettuce are the main ones; add endive, mustard-greens, cress, and chicory as chasers, and that's about the lot—at least in the sophisticated West. All of them can be lumped under that terrible phrase "green vegetables." "Eat up your green vegetables. They're good for you," was a command that most of us could well have done without. And yet there is more than a grain of truth in the statement that "they're good for you." Leaves—especially if prepared in the right way—can provide vitamins, protein, and fiber, all of which are essential to a balanced diet for *Homo sapiens*.

Left: a seemingly endless expanse of wheatfield—but this, and the many others, are still not nearly enough to feed an increasingly hungry world. We eat only the seeds of our annual crop of cereals, leaving much of the rest to rot. Such waste makes little sense today.

Right: take a look at any vegetable stand and you will see that much of the plant produce on sale is far from green. We eat fruits, seeds, stems, and roots in abundance, but even vegetarians eat very few actual green leaves.

There are, however, several problems. The first of these is that evolution left man with a great deficiency when it comes to obtaining the rich contents of plant cells. One of the main features that distinguish plants from animals is that every plant cell is surrounded not only by a cell membrane but also by a tough cell wall made of cellulose. Although cellulose is itself made of sugar molecules stuck together in the form of long, branched chains called *polymers*, they are stuck together so tightly that it is very difficult to tear them apart. The key that releases both the energy in the sugars that make up the cellulose and the proteins and other contents of the plant cell is a very special biochemical, an enzyme called *cellulase*. The digestive tract of human beings does not produce cellulase, nor for that matter, as we shall see, do the guts of any other higher animals. So although we have *proteases* for the digestion of proteins, *lipases* for the digestion of fats, and *aldolases* for the digestion of sugars, without cellulase we cannot open the box of goodies. This is a job for the decomposers.

Almost by definition, decomposers have to be able to break down cellulose; if they couldn't, the world would be tree deep in cast-off cellulose. Kings among the cellulose-crackers are the bacteria. Because none of the advanced herbivores produces its own cellulase, they rely on a good culture of bacteria in their digestive tract to open up the cell-sized larders. This is another example of the way in which organisms that we tend to put at the top of the evolutionary tree are absolutely dependent on organisms that are usually relegated to a very lowly position on the same tree. We human beings cannot take direct advantage of the friendly cellulose-crackers because we have a straight-through digestive system, and even with a good flora of bacteria in the gut, cellulose-cracking takes a long time. So the advice to "eat up your green vegetables" should really be "chew up your green vegetables," so that some of the cells at least can be broken down mechanically. In efficient herbivores, the gut is highly modified to permit a good measure of two-way traffic, so that partly digested food can be slopped back and forth in what are really a series of fermentation tanks, thermostatically controlled to allow the bacterial cellulases to work

For a long time, man has recognized his limitations when it comes to digesting vegetable matter and has taken steps to overcome the problem by pretreating such food—or having it pretreated by letting it be digested by a real herbivore first, and then eating the meat of the animal and drinking its milk. The quickest and probably most efficient method of pretreatment is either to cook vegetable matter or allow it to ferment before chewing. Even with modern advances in culinary technology, however, other difficulties present themselves, as we shall see—and so the bulk of the world's leaves still goes back into the soil unused by man.

Recently, there has been much experimentation

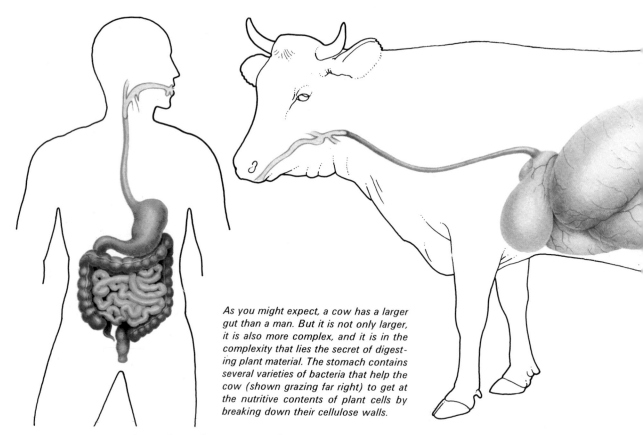

As you might expect, a cow has a larger gut than a man. But it is not only larger, it is also more complex, and it is in the complexity that lies the secret of digesting plant material. The stomach contains several varieties of bacteria that help the cow (shown grazing far right) to get at the nutritive contents of plant cells by breaking down their cellulose walls.

in an attempt to replace the ordinary cow with a machine that can produce milk from any leaves that are fed in the top. The product that flows from such a mechanical cow is a tolerable protein concentrate, but, although it looks and feels like milk, it tastes, not surprisingly, somewhat like vegetable juice. The potential of the machine would appear to be enormous when you consider that the protein content of all leaves is of roughly the same quality. One fact, however, limits the use of many leaves, even for feeding a mechanical cow: some varieties contain chemicals that are toxic, especially when concentrated. Just think of all the calories that are discarded with the leaf when you enjoy the succulent *petiole* (leaf-stalk) of rhubarb! Unfortunately, those leaves contain oxalic acid, which may be excellent for cleaning brassware but is lethal to human beings.

I have always been fascinated by the ability of plants to produce a store of poisonous substances in their living tissues without themselves suffering any obvious ill effects. The oxalic acid in rhubarb leaves is mild compared with the chemicals present in the plant juices that primitive tribes use on the tips of blow darts. The poisons spell instant death to their enemies—so why not to the plants

themselves? Most of the poisons are substances called *alkaloids*, of which the most infamous is strychnine, produced by the vomit nut (and no prizes for guessing how it got its name). Strychnine affects the nervous system and the muscles, with potentially catastrophic results. If administered in minute doses, it is a good stimulant—perhaps the best term would be "pick me up"—but an increased dose is even more effective as a "put you down." One reason why plants get away with such substances unharmed is that they are not complex organisms like animals, and lacking, as they do, either muscles or a nervous system, there is nothing on which the poison can work.

The best example of such a selective poison is the one produced by a small group of dino-flagellates. These small armored plants are important members of the marine plankton and are found in small numbers in most plankton samples. For reasons as yet not understood, they sometimes undergo what appear to be massive population explosions—so explosive, indeed, that the sea in which they are "blooming" becomes colored deep tomato red. A by-product of the life of this tiny plant is a very strong nerve poison. The dino-flagellates are an important food source for such

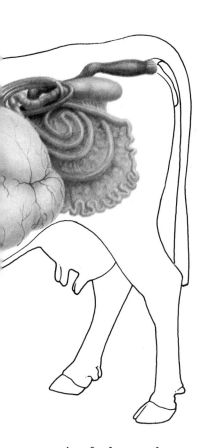

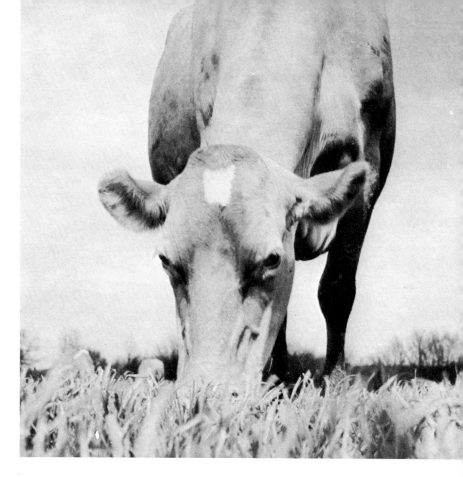

suspension feeders as the common mussel. When there is what is called a *red tide* the mussels gorge themselves with no ill result because their simple nervous systems are immune to the poison. But when a more complex creature such as man eats the dinoflagellate-stuffed mussels, the poison can go disastrously to work.

There is perhaps one other reason why plants can store poisonous substances and get away with it: the majority of plant cells have *vacuoles*—internal spaces, each surrounded by a complex double membrane called the *tonoplast*. The vacuole can be used as a convenient trash-can; substances produced but not required by the plant, including poisons, can be safely filed away within the tonoplast. We do not know exactly how this double-wrapping system works but it certainly does work.

Not all the alkaloids are as dramatically harmful in small doses as strychnine. A refreshing cup of tea or coffee is loaded with caffeine. Quinine (familiar to us as an ingredient of tonic water) has kept many an explorer on his feet in malaria-infested tropics. And everybody knows that nicotine does not kill—at least, if taken in moderation, not right away. Basically, though, it is best to be wary of all alkaloids. Even the more

innocent ones can hurt human beings if not used with care and discretion. And with some of the others, the best possible slogan is: "The plant may get away with it, but we shan't."

Aside from the fact that toxic chemicals are present in a good many leaves, however, there would seem to remain numerous possibilities for nourishment that we have not yet drawn upon. Mahatma Gandhi once said: "The unlimited capacity of the plant world to sustain man at his highest is a region as yet unexplored by modern science. I submit that the scientists have not yet explored the hidden possibilities of the innumerable seeds, leaves, and fruits for giving the fullest possible nutrition to mankind." This statement is reinforced by the knowledge that the flora of four fifths of the world's surface has not as yet been fully investigated. We have hardly started to catalog the products of evolution, and plant taxonomy often stands at the bottom of the table of importance when it comes to teaching and research within the plant sciences. Our current acute need for massive increases in food production, however, is at last galvanizing scientists into bringing all the most sophisticated scientific techniques to bear on this mammoth problem.

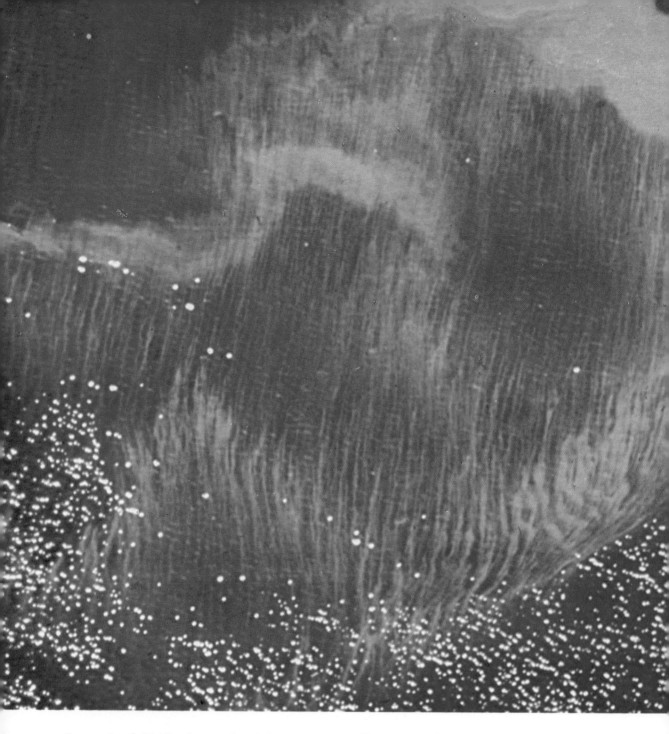

One major field of endeavor that is beginning to pay off aims at developing plants that are more efficient at converting the incident energy of the sun into potential food. As a plant grows it puts on more leaves, and—in theory, at least—more leaves should intercept more light. But even with a very efficient leaf mosaic, the older leaves at the bottom of a plant will eventually be shaded so much that their photosynthesis is impaired and they use up as much energy as they can fix.

From that point on, they begin to put a drain on the total photosynthesis of the plant. The easiest way to measure these functional aspects of a plant system is to measure its *leaf-area index* (that is, the area of leaves as compared with the area of soil they cover). From the point when the older leaves become a drag on the system—we call this point the *optimum leaf-area index*—onward, the rate of growth must begin to slow down. As more leaves are produced, the situation

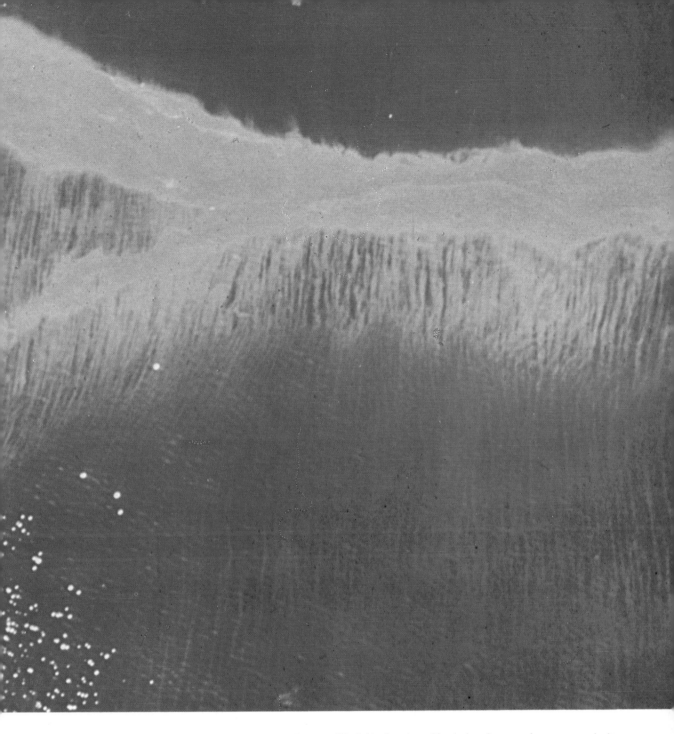

at the base of the plant will get critical in that the basal leaves start to die off, and at last the rate of die-off at the bottom will be exactly balanced by the development of new leaves at the top. Once the leaf-area index has become static, the plant will soon stop growing.

Expressing the whole thing as leaf-area index gives an insight into the performance of the crop system in relation to the environment. And the more we learn about this relationship, the more

"Red tides" such as this one in a Japanese bay are caused when, for natural reasons not yet fully understood, there is a population explosion among certain species of the tiny floating plants called dinoflagellates. Although microscopic, they are present in sufficient quantity to color the sea with reddish streaks. The tiny plants produce minute quantities of a poison that is harmless until concentrated by animals feeding on them. Thus red tides are a serious threat to man, who may eat filter-feeding organisms, such as mussels, that concentrate the dinoflagellate poison within them, but are themselves completely unharmed by it.

Above: belladonna means "lovely lady," a fitting name for a plant that has beautiful bell-shaped flowers and shining black berries. But this plant is also commonly known as deadly nightshade, because it contains the poisonous substance called atropine.

we should be able to improve the ability of plants to grow productively. This is of enormous importance to farmers, whose livelihood depends on using the right crops to manipulate the potential of their land. Ideally, the farmer wants to exploit the most productive part of the crop-growth cycle, in other words to bring a given crop only to the point where leaf growth is at a maximum. The reason for this is obvious: after that point the farmer's investment in soil gives progressively

Left: a young rhubarb plant. Rhubarb is a plant whose leaves should not be eaten. The succulent red leaf stalks are delicious, but the leaves are full of poisonous oxalic acid.

less return, at least in terms of total organic production. In many crops, only the fruits and seeds are marketable, and yet the plants must be left in the ground long after they have borne fruit. Thus the farmer as well as the consumer should, in theory, benefit from any scientific effort to learn how to make better use of leaf production. "Let us eat up our green leaves" might well be a very good slogan for the future of mankind.

Natural ecosystems are not subject to the same limitations of leaf area as farmland, because in the case of natural ecosystems not just one but many plants make up the canopy. Each plant can make use of the various qualities and intensities of light within the area. The diverse community is structured in both space and time, and thus partly overcomes the limitations of leaf-area index. Here, as so often, nature has much to teach us, if we are willing to learn. It is becoming possible for modern farmers to increase the potential of

The darts used in the blowpipe of this Borneo hunter are tipped with deadly plant juices. The earliest classification of plants was into two groups: those that do good, and those that do harm. Man soon learned to use both sorts to his advantage.

The leaves of these purple foxgloves produce a substance used for treating heart disease, but also contain a noxious poison.

their farm systems by including a number of crops in one growing season, and even by turning to polyculture (that is, growing more than one crop at a time). In the future, as our technology relating to leaf protein develops, plants that have always been thought of as weeds may well become an important by-product of the farm system. In addition to enlarging our food supply, such an advance could save an enormous sum of money on herbicides and hand weeding.

The example has already been set, at least in

Caffeine and nicotine, both substances extracted from plants, can do great harm to man if they are taken in immoderate doses. The difficult question is: What constitutes moderation?

part, by such crops as oil seed (castor oil, palm oil, and others). In the early 1960s, some 80 million tons of oil seeds were produced in the so-called "developing" nations of the hotter parts of the world. After the oil had been extracted, the residue of the crop was destroyed. That "useless" residue contained 15 million tons of protein— one fifth of the amount required to supply the basic need of every living human being in the world. To make this type of protein useful to man involved all sorts of problems, but scientific research solved them. Today the residue is worth as much as the oil on the open market, especially in America, where it is used in the manufacture of pig and poultry foods. It is no doubt ironical that a rich source of protein produced in countries where the population suffers from protein deficiency helps to feed those in the more affluent countries. But that is a problem of economics, not of the plant sciences, which are making great advances toward a more efficient use of the land.

One of the greatest success stories to date is that of the green pea, which, thanks to the technology of deep-freezing, has in many countries been turned from a summertime treat into a year-round staple. In response to developments in the field of refrigeration engineering, plant scientists have recently produced the "super pea" for farmers' fields. The type of wild pea that started the whole thing off came from Ethiopia. It was a rather untidy-looking plant, which straggled all over the place, finding support wherever it could, and producing on each branch a single pod with a few very small peas. These tiny peas were the important part of the plant, for each contained a supply of protein and energy that could be made readily available to man by cooking. What was needed was a super-pea plant and a super-crop system that, together, could in the short pea season provide enough peas to fill the freezers and keep a large population going for a whole year. The stakes were high, and so all the guns of the plant and agricultural sciences were trained on the problem.

By now the pea plant has been shaped into a more orderly form, productive in every way, with leaves and branches arranged to intercept the sun more efficiently. There are two pods on every branch, each pod bursting with succulent green peas, not too big, not too small, but just right for the freezing process. And just as the plant breeders have improved on the naturally evolved product, which is still best called the "Mark 1 Ethiopian

Wild Pea," so too, the agricultural sciences, farmers, and frozen-food firms have each done their bit. Every farmer within a specific radius of a freezer plant contracts to plant a certain acreage of peas on a certain day, so that they will be ready for harvest on a known date. The date is of enormous importance, because the peas are at their best for freezing for only about 24 hours. By absolute programming of the crop, the expensive and complex harvesting machinery can be kept employed full time. It is a case of "all systems must be ready to go" if the whole man-made mechanized complex that is the frozen-pea ecosystem is to be kept in a viable state.

Evolution of the modern frozen-pea ecosystem is not finished yet. One of the main hang-ups in the system used to be the question of what to do about the protein-rich leaves that clog up the great pea harvesters. Because both the stems and the pods of the plant contain chlorophyll and are well supplied with stomata, the leaves of pea plants are obsolete; and so breeders have just produced the leafless pea—in which problems of leaf-area index are replaced by those of pod-area index.

There are, of course, other problems related to the traditional monoculture (cultivation of a single crop per field) of most farmers. The main problem is that such crops are excessively susceptible to the catastrophic effects of disease. Once a disease gets a foothold, it can run through a single crop much more easily than it would through a diverse natural community. For the most part, this problem has been sidestepped by the fact that most of our crop plants are grown in parts of the world far removed from their point of origin, and thus many of the diseases have, as it were, been left behind. Sometimes they do catch up, though, especially in these days of jet travel, when planes may carry not only vacationers but spores of disease-producing organisms. In fact, as the 20th century has progressed, geographical isolation has more and more become a thing of the past. Plant breeders must therefore be on the watch for diseases, and one of their main tasks is to build disease resistance into their new super-crops. Unfortunately, it is not just the crops that are evolving under the manipulative power of the breeder; disease organisms are themselves evolving, with no help from anyone. And so the breeding game involves a constant effort on the breeder's part to keep a couple of jumps ahead in what appears to be a never-ending race between better crops and more potent disease organisms.

Then, too, most of the success stories of both intensive and extensive agriculture still come from the temperate zones, not from the tropics. This seems unfortunate in view of the fact that the humid tropics hold the greatest potential for plant production. Has the imbalance occurred because the money for experimentation has always—at least until recently—been spent in the developed countries? Or is it that there is some limitation to tropical agriculture that we do not yet understand? And one final essential

More exposed leaf area means more photosynthesis. Plant breeders hope to develop plants with better leaf arrangements so that the upper leaves do not deprive the lower ones of light.

question: Is there, after all, any real validity to the concept of "success stories"? That last sentence may sound strange, but it is one we must face up to. If an overall balance sheet for long-term agricultural success in all the regional climates of the world were drawn up, *would* the tropics really come out worse than Europe or North America? Central Africa, India, and Southeast Asia all have extensive histories of agricultural systems that have supported large populations for a long period of time—much longer than some of the instant success stories of the temperate zones. Add to this the fact that there are already signs that the new mechanized kind of agriculture may be having adverse effects on the soil structure and on the natural balance of the landscapes in which it has been developed. A question now being insistently posed is this one: How long can we continue to "abuse" the system, without bringing on some massive catastrophe? Would the use of more leaves as a direct source of protein, or even breeding of more leafless plants, not hasten the rate of soil deterioration? As more and more of the world's people are coming to depend on the productivity of the extensive systems of temperate agriculture, the right answer to such questions may be the key to the survival or extinction of man.

Still, we must—and do—go on trying. Much effort is at present being invested in the tropics, in an attempt to cash in on what is still reckoned to be latent potential. The bulk of the results indicate that the best way to increase usable plant production is for farmers to make use of small units, managed by careful rotation of crops, rather than to develop massive crop systems. Although such small units require intensive labor, they at least have the protection of built-in diversity. Up to now, the majority of the world's catastrophic famines have occurred in the tropics; and they have not been caused by a breakdown of indigenous small-unit farming systems, but by long periods of bad weather, especially drought. The classic account of famine in Pearl Buck's prize-winning book *The Good Earth* tells in detail of the sufferings of a group of Chinese peasants. It also shows, though, that these "ignorant" peasants had the know-how to derive at least some benefit from their land even during famine, and were thus able to ride out the lean years. It is instructive to compare what happened when famine hit a much more sophisticated agricultural system based on one crop, the potato.

The climate of Ireland is so suited to raising potatoes that, after their introduction from South America in the 1500s, they took over from cereals as the staple of the peasantry. In fact, the potato was so successful that the population doubled in the 50 years from 1795 to 1845. Then a mold called *late blight* spread with devastating effect through the great monoculture. Ireland's population was slashed in 15 years from 8 million to its pre-potato level of about 4½ million. One and a half million died of starvation and related diseases, and roughly the same number emigrated to North America. It is tempting to suggest that maybe the natural carrying capacity of the Emerald Isle (in terms of omnivorous human beings) is no more than about 4½ million—its current population as well as that of 100-odd years ago. Again and again, the evidence shows that the soil/climate system of any given area is capable of being productive at a certain "evolved" level for a long time, but there are limits to how high that level can go. The system may be forced to yield a higher return, but only for a time.

As we approach the end of the world's stocks of

Quick freezing has turned the traditional green pea (above left) from a summertime treat into a year-round vegetable. Plant breeders have now developed a new variety (shown above right in an unripe state) that grows without wasteful leaves; there is enough chlorophyll in the stems and pods of the developing plant to keep it supplied with energy, and leaves are therefore unnecessary.

coal and oil—the affluence of past ecosystems and the deposit account of world energy—we must look to the only dependable and lasting wealth we possess: the sum total of current photosynthesis. Even if we can overcome the problem of harnessing atomic power, or can find some other source of energy that does not depend on green plants, the annual production of organic matter will remain the mainstay of all our synthetic industries. Plants are here to stay—and not only for food! We are just beginning to make the measurements that allow us to understand how the productive natural systems work. The ecosystems of the world are the workshops of evolution, and there is no other place in which we can study the process. If any part of the world matrix of vegetation or any component of any vegetation type is destroyed, part of the necessary information has gone, and gone for ever. Conservation, then, is the *sensible* utilization of all natural resources, most important of which is the world of plants.

Only if we understand the world of plants can we continue to evolve a viable human society. There is little doubt that evolution works through mutation to produce new structures, and through natural selection to test their adaptability and survival value. Where there is doubt, though, is in the interpretation of the dogma of "Survival of the Fittest." If this is the true philosophy of evolution, it gives *carte blanche* for the rat-race philosophy of man—one that can end only in catastrophe, and that will include King Rat at the top. The true philosophy of evolution envisions the development of integrated organic systems, each member and each part of which are of equal importance. Together, but only together, all forms of plant and animal life can reap the full potential of the environments of this earth.

Pineapple fields in Hawaii. The organized layout of this plantation reflects modern scientific planning for the best possible utilization of resources. For example, each field is twice as wide as the boom of the mechanized spraying machine and of the harvester's conveyor belt; and its corners are rounded to fit the turning circles of the machines. Such planning saves both time and energy.

138

Index

Page numbers in *italics* refer to illustrations or captions to illustrations.

Picture Credits

Cover: Heather Angel
Title page: G. Rönn/Östman Agency
Contents: G. R. Roberts, Nelson,
    New Zealand
9   Jacana
10   Icelandic Photo © Mats Wibe
    Lund
11   Heather Angel
12, 13   Gordon F. Leedale/
(TR, BL)   Biophoto Associates
13(BR)   Eric V. Grave/Photo
    Researchers Inc.
14   Hermann Eisenbeiss, München
15(TR)   Eric V. Grave/Photo
    Researchers Inc.
15(BR)   Dr. Walker (Banta)/N.H.P.A.
16(R)   © Douglas P. Wilson
17   John Markham/Bruce Coleman
    Ltd.
18(L)   Heather Angel
19   John D. Dodge, Birkbeck
    College, London
20(L)   Heather Angel
21–2   Josef Muench
24   David J. Bellamy
26–7   Heather Angel
28   Thase Daniel/Bruce Coleman
    Inc.
29(R)   Ken Brate/Photo Researchers
    Inc.
31   G. R. Roberts, Nelson, New
    Zealand
33   Heather Angel
36–9   Heather Angel
40(L)   M. Savonius/N.H.P.A.
41(L)   Frederick Ayer III/Photo
    Researchers Inc.
41(R)   M. Chinery/Natural Science
    Photos
42   Heather Angel
43   Norman Myers/Bruce Coleman
    Inc.
45   Edward S. Ross
46(L)   Heather Angel
47   Picturepoint, London
48, 50   Heather Angel
51(TR)   Heather Angel
51(B)   David J. Bellamy
53   G. Ronald Austing/Bruce
    Coleman Inc.
54(L)   Heather Angel

55   Charles Sinker
56–7   John Markham/Bruce Coleman
    Ltd.
58(TL)   Edward S. Ross
58(BL)   C. Ott/Bruce Coleman Ltd.
59   Jane Burton/Bruce Coleman
    Ltd.
60   Heather Angel
61(R)   G. A. Matthews/Natural Science
    Photos
62   Anthony Huxley
63   Heather Angel
64   Emil Schulthess
65   Heather Angel
70–1   Fred Bruemmer
72   David J. Bellamy
73   Fred Bruemmer
74   S-E. Hedin/Östman Agency
75   Nick Impenna/Photo
    Researchers Inc.
76   J. & D. Bartlett/Bruce Coleman
    Ltd.
77   Hayon/Pitch
78(L)   M. E. Warren/Photo
    Researchers Inc.
78(R)   Dr. W. M. M. Harlow/Photo
    Researchers Inc.
80   Edward S. Ross
81(R)   Aldus Archives
82   Jacana
83   Dr. Ivan Polunin/N.H.P.A.
86–7   Heather Angel
88   Dr. Ivan Polunin/N.H.P.A.
90(L)   Heather Angel
91(T)   Dr. Ivan Polunin/N.H.P.A.
91(B)   L. West/Frank W. Lane
92(T)   Heather Angel
92(B)   G. A. Matthews/Natural Science
    Photos
93(R)   Dr. Ivan Polunin/N.H.P.A.
95   Fred Mayer/The John
    Hillelson Agency
98(L)   Picturepoint, London
99   Douglas Botting
100   M. Fogden/Bruce Coleman Ltd.
101   David J. Bellamy
102   Farrell Grehan/Photo
    Researchers Inc.
103   Heather Angel
104   Virginia Carleton/Photo
    Researchers Inc.

107–8   Josef Muench
109   G. R. Roberts, Nelson, New
    Zealand
110   Anthony Huxley
111(TL)   J. & D. Bartlett/Bruce Coleman
    Ltd.
111(R)   Russ Kinne/Photo Researchers
    Inc.
113(T)   Jack Dermid/Bruce Coleman
    Ltd.
113(B)   Peuriot/Pitch
114   Heather Angel
115   David J. Bellamy
116   G. R. Roberts, Nelson, New
    Zealand
117   Heather Angel
118-9   David J. Bellamy
120   Dr. E. Gelpi/Östman Agency
123   Stephen Dalton/N.H.P.A.
124(BL)   Bruce Coleman Limited
124(BR),   Jane Burton/Bruce Coleman
125   Ltd.
126   United States Information
    Service, London
127   Michael St. Maur Sheil/Susan
    Griggs Agency
130–1   Georg Gerster/The John
    Hillelson Agency
132(TL)   Heather Angel
132(BL)   Bruce Coleman Ltd.
133(L)   J. Alex Langley/Aspect
133(R),   Heather Angel
135   Heather Angel
136   A-Z Botanical Collection Ltd.
137(R)   John Innes Institute, Norwich
138–9   Georg Gerster/The John
    Hillelson Agency

Artist Credits

© Aldus Books: (Photos) Mike Busselle
124(T), 134; Geoffrey Drury 129(R); (Artists)
Henry Barnett 25; Hatton Studio 16(L),
96–7; Amaryllis May 96–7; Sean Milne
128; David Nockels 30, 52, 68–9, 84–5;
Joyce Tuhill 34, 97

© Geographical Projects Limited, London
67

GREEN WORLDS

Part 2
Forest Life

by Michael Boorer

Series Coordinator   Geoffrey Rogers
Art Director   Frank Fry
Design Consultant   Guenther Radtke
Editorial Consultant   Malcolm Ross-Macdonald
Assistant Editors   Bridget Gibbs
Allyson Fawcett
Copy Editor   Damian Grint
Research   Naomi Narod
Carol Potter
Art Assistants   Amaryllis May
Michael Turner

Contents : Part 2

Editorial Advisers

MATTHEW BRENNAN, ED.D. Director, Brentree Environmental Center, Professor of Conservation Education, Pennsylvania State University.

PHYLLIS BUSCH, ED.D. Author, Science Teacher, and Consultant in Environmental Education.

MICHAEL HASSELL, B.A., M.A.(OXON), D.PHIL. Lecturer in Ecology, Imperial College, London.

STUART MCNEILL, B.SC., PH.D. Lecturer in Ecology, Imperial College, London.

JAMES OLIVER, PH.D. Director of the New York Aquarium, former Director of the American Museum of Natural History, former Director of the New York Zoological Park, formerly Professor of Zoology, University of Florida.

MICHAEL TWEEDIE, M.A. formerly Director of the Raffles Museum, Singapore.

Introduction

The high living standard currently enjoyed by part of the human race has been achieved largely at the expense of the world's forests, past and present. Whether it takes the form of modern teak furniture or a box of matches, the forests supply our timber. The page on which these words are printed is made of wood pulp. The forests were the first source of the rubber for our tires, and the forests of long ago built up the stock of oils that both lubricate our automobiles and power them on their way.

While affluent man speeds heedlessly toward the mirage-goal of even greater affluence, it may well be that he is really heading toward ecological disaster. Until the present time the forests have continued to meet our ever-increasing demands, but there are ominous signs of danger ahead. The oil is running low, and the living forests are dwindling. If any complex machine or system is to keep running, all of its parts must work together without strain. Ancient man was at one with his environment. He hunted in the forests and gathered fruits, but used up natural resources no faster than they could be replaced. Modern man is still part of his own total environment but, inspired by intelligence and driven by greed, he is no longer in step with it. This is why it is essential that we should understand the ecological system of which we are part. With understanding and well-directed intelligence we must, in the very near future, make the effort to gear our life-style to that of the world about us.

The forests represent the most advanced state that wildlife on land can attain. Their living parts—always interesting and often beautiful—all work together in such a way that potentially the whole living system is indestructible. For those who are willing to understand, the forests have much to teach.

The Magic of Forests

For most of us there is something very special about forests. In our imagination they are the least tamed, the most secret, and at the same time the most inviting of the earth's wild places. The bleak pine forests of northern lands, the leafy woodlands of temperate climates, and the torrid jungles of the tropics have an almost magical appeal. Within their shade, with the branches interlaced overhead and the trunks of the trees closing in on all sides, there are no distant horizons; we are compelled to concentrate on the abundant life close at hand. The forests form intimate and yet teeming worlds of their own.

Once, not so many generations ago, the parts of North America and Western Europe where so many of us live our busy, urban lives, were forest-covered. Our ancestors felled the trees for fuel and building materials, and cleared open spaces for agriculture. To them, the forests represented untamed nature and freedom from workday cares. In England, Robin Hood sought freedom in the greenwood of Sherwood Forest, and the forest itself adds to the power of his legend. In America, the log cabin in which Abraham Lincoln was raised was on the edge of a forested wilderness, and just such a wilderness formed the wild frontier that Davy Crockett knew.

Perhaps most of us subconsciously feel that we were born too late, and yearn for the freedom that our vanished forests represented. Certainly they can affect us powerfully, and men of influence have always sought to have forests of their own. The Norman Kings of England enclosed the primeval New Forest for their own sport and pleasure, and powerful men born after the forests had disappeared sought to re-create them. In more modern times, the trees of city parks and the mightier stands of timber of our national forests and parks are the "private" forests of everyman. On our afternoons off, at weekends,

Beavers once lived in all the rivers of the forests of Europe (including Britain), northern Asia, and North America. They felled trees as food and as building material for the dams that they built in order to maintain a constant water level. They made a visible impact on the forest environment, yet were in balance with it. Our ancestors knew the beaver well, but now it has gone from much of its former range.

8

and on vacation these parks lend a touch of forest magic to our lives. With their aid, for a moment at least, we can all return to the wild and be free from care.

Even the most weighty and impersonal works of forest science are not immune to this magic. The forests mean so much to us that the memories come unbidden, and we are carried away upon the wings of imagination. While reading a learned paper on the larval stages of one of the more obscure insects of the Scandinavian forests it is impossible to empty the mind of other images; it only needs the slightest lapse of concentration for the memories to come crowding back—the vast stands of soaring pines and firs, the sighing of the wind among the pine needles, the bittersweet smell of the resin, and the softness of the dead pine needles underfoot.

Each type of forest has its own particular charms to haunt the memory. In the broadleaved forests of temperate lands the cycle of the season adds never-ending variety. It is almost impossible to choose a favorite season, but for many people the fall casts the strongest spell. It is a time of such fruitfulness, and such incredible color. When the gales roar through the branches, setting them tossing and dancing, and the yellow, golden, and russet-red leaves come spiraling down to carpet the ground, turning brown, and rustling crisply at each stride, then the sounds and the smells of the fall—that soft smell of fallen leaves—are without comparison.

For most of us the lush forests of the tropics are far away, and perhaps for this reason the most romantic of all. Nowhere else on earth is plant life so luxuriant. Giant trees reach up until their crowns are invisible from the ground, lost among the crowded canopy of leaves overhead. The thick stems of vines also climb upward, emphasizing that forests, unlike deserts and grasslands have a third dimension. They have depth as well as length and breadth. To walk in a forest is almost like walking at the bottom of a leafy sea, and in the tropics the dampness and the humidity add to this effect. Some animals of the tropical forests—the leeches and the giant flatworms—belong to water-loving groups, and heighten the impression that we are submerged.

The animals of the world's forests are shy and unobtrusive, but they are a vital part of the forests' lure. An English professor of education studied the questions that young children asked their mothers, and found that many of their questions were about animals. One that he noted was "Mummy, are there bears in that wood?" For most children, and for most adults too, bears have a stronger appeal than flowers. The animal life of the forests is subtle yet pervasive. The busy lives of the ants among the leaf-litter, and the incessant search for insects by the small birds among the branches can easily pass unnoticed. Often the first and only sign of the larger animals is the sound of them crashing through distant undergrowth as they make their escape, startled by our intrusion. How often, seeing but unseen, they must watch us pass by.

Most of us are townsmen, and have lost the woodcraft skills that our ancestors once had. We need the dwindling forests, for our lives have become so hectic that never before has recreation—in the full meaning of the word—been so important, but in order to see them clearly we need an unobtrusive guide. In the following pages the aim is to capture in words and pictures the magic of the forests and their animals. With the aid of modern science, and especially that of *ecology*—the study of the way in which the plants and animals act and interact in living communities—it is possible to understand the world's forests more clearly than ever before.

When they are not disturbed by man, plants grow, produce seeds, and die in an unending succession. There is a silent and unceasing struggle between individuals and among species for living space, for sunlight, water, and nutrients from the soil. Only the driest and coldest parts of the earth's surface and the darkest depths of the sea fail to provide conditions suitable for plants of one kind or another. In the course of time every other part of the world has become covered by those kinds of plants that can best thrive in the conditions that are available. Where grasses flourish the result is prairie, steppe, or veldt, but where the rainfall is higher, trees become established and grow upward, catching most of the available sunlight so that little is left for other types of vegetation.

The relationship between plants and their environment is not a simple, one-way affair. The

As spring comes to Sweden, the deciduous trees of the forests spread their broad new leaves to catch the energy that radiates from the sun. The evergreen conifers have kept their leaves throughout the winter. In forests all trees compete for the light that is available, and there is more than one way of obtaining it.

leaves of green plants are spread to catch the energy that radiates from the sun, 93 million miles away. Using this energy, they build up sugars from the carbon dioxide of the air, and the water from the soil. The sugars, in turn, combined with mineral nutrients obtained from the soil by the plants' extensive root systems, are used to build the materials of which the various parts of plants, such as leaves and branches, are made. All this activity does more than just build plants —it makes soil, too, for as the plants die they decay and the materials of which they were formed are returned to the soil. The richer the soil becomes, the better able it is to support the growth of trees, so there is a steady trend toward the development of forests if rainfall is sufficient.

There are many kinds of forests, each with its own typical mixture of trees, shrubs, and smaller plants—although sometimes a forest of one kind merges almost imperceptibly with another. However, it is possible to distinguish forests of three main types. In colder parts of the world conifers predominate. Their needle-shaped leaves conserve water at all times, a quality that is especially important when the ground is frozen. Because the leaves are not shed all at once, they make the most effective use of the light that is available. In temperate climates most trees have broad leaves that are lost in the fall in order to minimize frost damage. In the tropics most of the trees have broad leaves at all seasons, for the tropical forests grow in regions where there is no seasonal shortage of water and frost does not occur. By shedding their leaves only a few at a time the trees are able to take advantage of the sunlight all the year around.

Forests offer many ways of living—or to use the biologists' term, ecological niches—for many kinds of animals. There is living space among the branches, on the ground, and in the surface layers of the soil. Sap, bark, buds, leaves, flowers, fruits, and seeds all provide food for *herbivorous* or "plant-eating" animals, and these in their turn are the food of *carnivorous*, or "meat-eating" animals.

During the long course of evolution the animals of the forest have become progressively better adapted for climbing, clinging, and leaping, or perhaps for running on the ground beneath overhanging branches. They have been shaped by their environment, and in their turn they help to shape it. For example, plants that were once ideal as food for animals have become adapted for survival by the development of protective spines, or a bitter, unpleasant taste. Other plants have become adapted in ways that make use of the animals present; some trees produce seeds contained in fleshy fruits that serve as tasty bribes to animals that consequently distribute the seeds. Not all plant material is so palatable; in fact much of it is often tough and difficult to digest. Animals that feed on it need an efficient grinding mechanism to break it down into small pieces before digestion even begins. Herbivorous mammals, for example, have premolar and molar teeth with ridged surfaces near the back of the mouth. These act rather like mill-stones.

Many of the smaller herbivorous mammals are rodents, such as mice, squirrels, and porcupines. Their chisel-like front teeth never stop growing and actually become sharper with wear. They are ideal for opening nuts, or for stripping the bark from trees. Some of the larger herbivorous mammals sometimes feed on grass and other plants growing under the trees, a process known as grazing. A bigger group feeds by stripping the branches, leaving only the wood, which no mammal can digest. This method of feeding is called browsing. The grazers usually have square, firm lips, but browsers have more pointed, movable lips, ideal for wrapping around branches and for tucking them into their mouths.

Even when plants die they still continue to provide food for some animals. Some beetle larvae, for example, feed on dead wood. Alternatively the dead parts of plants are consumed by the fungi and bacteria that cause decay. In this way their chemicals again become available to be used by other living plants. In the wild there is no waste. Animals that eat plants are, in turn, eaten by a host of carnivorous animals. Centipedes feed on grubs, small worms, and snails among crevices in the bark, while small birds search the leaves and branches for insects. The larger herbivores are the food of carnivorous mammals that lurk and pounce, or chase after their prey. In defense against these attacks herbivores use a wide variety of tactics. Some tend to keep still and make use of superb camouflage, while others rely on sheer speed to escape.

Many forest animals have taken to the third dimension—the vertical one—that the forest offers. For the very small ones especially, climbing among the branches presents no great problem. All they need is suction pads on their feet, or claws to dig into the bark. Larger climbers

Above: many climbers have long tails, which they use in balancing, and some of them have tails that are prehensile—able to grip the branches. This giant skink from the Solomon Islands is supporting itself with its hind limbs and its tail.

The American black bear is one of the largest of climbing mammals. It grips the bark with the leathery pads on its palms and soles, and uses its claws as climbing irons.

Within the three-dimensional world of the forests it is very useful to be able to fly, but all fliers must rest at times. Bats cling by means of their fingers and toes.

The Cuban tree frog can cling to smooth branches and leaves with suction pads on its fingers and toes. Its small size is another adaptation to its life style.

Below: the eucalyptus-loving Australian koala climbs by holding on with its fingers and toes, which are equipped with sharp claws. When it grips a branch two fingers oppose the other three, so that the hand grips firmly. In other climbing mammals, such as monkeys, the thumb opposes the other four fingers.

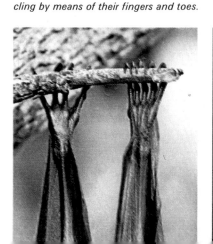

Like many of the larger mammals of the forests, the leopard is patterned with spots that blend well with leafy shadows. Camouflage is essential to both hunters and hunted.

Right: all true chameleons can change color to match their surroundings. This species is particularly well camouflaged, being leaflike in outline, and having a stripe that resembles the midrib of a leaf.

Below: camouflage is important at all stages of an animal's life history. These young owls are most vulnerable when they first emerge from the nest, and at this time their neutral coloring helps to make them inconspicuous.

Above: being the natural prey of keen-eyed birds, insects have a special need of good camouflage. This bug from Brazil is leaflike in both shape and color.

Below: the patterns on the wings of the peppered moth closely resemble those made by lichens on tree trunks.

often use claws too, and some have long toes that can encircle and grip the branches. If one toe can oppose the other—as the human thumb opposes the fingers—the result is a very secure clamp.

Balance is important for large climbers, and here a long tail is useful. Sometimes it, too, can be used for clinging. The position of the eyes is also important. An animal can judge distance much better with two eyes instead of one, so the best climbers usually have both eyes on the front of their heads, instead of one on each side. Flying animals can move deftly from tree to tree and reach even the topmost branches with ease and, not surprisingly, birds and winged insects are very numerous in forests.

In fact it is the birds, the butterflies, the monkeys, and the squirrels—the animals that live among the branches and flit or scramble lithely from one to another—that are first noticed by even the most casual stroller through a forest and that usually provide his most enduring memory of it. The creatures of the forest floor may never be more than a distant rustle in the undergrowth but it would indeed be an unlucky walker who saw no bird or marten or monkey or whatever tree dwellers the forest might shelter.

It takes patience, and dedication—one might even say love—to discover the *other* life of the forest, the secret life mentioned earlier: the silent battle among the plants; the lives and habits of the shyer and smaller creatures; the damp dark world below the litter where there are struggles and relationships as intense and as fascinating as any in the lighter world above. These are secrets the forest yields slowly. To unravel them, and thus to learn something of the vast interwoven skein of life below, among, and above the trees, is one of life's great adventures.

The harpy eagle of tropical America is predominantly brown in color. This provides serviceable camouflage under a wide variety of conditions. It is for this reason that modern soldiers' uniforms are usually brown.

Right: there is more than one way of hiding from danger. The common mole of the woodlands of Europe and western Asia spurns camouflage, spending almost all of its time hidden underground.

Coniferous Forests

The northern coniferous forests encircle the world, stretching across North America, Europe, and Asia in an almost continuous belt. In each of these continents they are very much the same, often sharing the same species of plants and animals. This is not surprising since, in the not-too-distant past, the narrow stretch of sea between North America and Asia known as the Bering Strait did not exist, and the northern continents were joined. There was thus no water barrier to prevent land animals from moving across what are now separate continents.

In the cool north there is plenty of water—in fact the soil is often waterlogged—but for much of the year that water is unobtainable, locked away in the form of ice or *permafrost* (permanently frozen subsoil). At such times it is vital that plants should not lose water they cannot replace. For this reason the leaves of conifers have a thick cuticle, or outer covering, which cuts down evaporation and which also protects them from the effects of frost. In addition, their needle shape provides little support for snow, and does much to protect them from this for there is the danger that the branches will break under the weight of collected snow. In mid-winter, when the temperature may remain below freezing for weeks on end, the branches may bend beneath the snow's weight, but they rarely break. However, even when they do break conifers have a marvelous defense against would-be invaders or profiteers, from bacteria to large herbivores.

The needles, bark, and timber of conifers are rich in resin, which, if the tree is damaged, flows out to seal off the wound and, acting rather like a scab of dry blood, keeps harmful bacteria and fungi at bay. The resin's strong and rather bitter flavor deters many animals that would otherwise feed on the trees.

Most conifers are evergreen, keeping at least some of their leaves throughout the year. Only a few kinds, such as the larch, are deciduous, losing

During the icy grip of winter the coniferous forests of the north provide a bleak and forbidding environment. At this season many of the forests' animals hibernate or migrate, but some remain active, feeding hungrily on whatever food they can find.

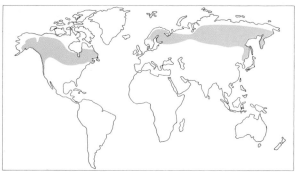

The World's Coniferous Forests

The map at left shows the areas of the world's coniferous forests. At the foot of this page are drawings of trees typical of the coniferous forests of eastern North America, to the same scale as similar drawings on pages 44 and 82. Below the trees are details of their foliage. On the facing page shafts of sunlight penetrate the canopy of a coniferous evergreen forest to reach the sparse undergrowth of the forest floor. The absence of a shrub layer is evident. Most of the trees in coniferous forests are evergreen. Often trees of a single species cover wide areas and so there is only a single-layer canopy. Because of this, and because of the sparseness of the ground-level vegetation, there is only a limited variety of living-spaces for animals in these forests.

their leaves in the fall. In the north, where the sunlight is never as intense as it can be in other parts of the world, trees that keep their leaves all the year around are best able to make full use of whatever light is available, but even the leaves of evergreens must ultimately fall. The floor of a coniferous forest is thickly carpeted with dead needles that decay only slowly because the resin they contain acts as a preservative. In the dense shade beneath the trees there is little light for undergrowth or for broadleaved plants such as birch or alder that are adapted to cold and to northern lighting. Only in clearings, or beside the many lakes and streams can such plants flourish.

The world's coniferous forests are not found only in the north. On the mountains in parts of the world farther to the south conditions are very similar to those of the northern forests. It is for this reason that parts of the slopes of the Rockies of North America, the Alps and the Pyrenees of Europe, and the Ural and Altai mountains of Asia are pine-clad. These coniferous forests of the

mountains are very much like those of the north. Their environment, too, is cold and undergoes marked seasonal changes. Predictably, they support similar kinds of plant and animal life.

The coniferous forests are a difficult environment to live in, for although they are often as warm as the deciduous forests to the south during the summer, they are bitterly cold in the winter months. Also, in northerly regions there is a considerable difference between the length of summer and winter days. Even during the long days of summer, the sunlight falls obliquely, providing less energy for growing plants than it does nearer the equator. The animals of the forests survive only because they are perfectly adapted to take full advantage of the summer, and to survive the winter.

Summer is the time of plenty. It is the season when the trees grow, providing food for many insects, birds, rodents, deer, and other herbivores. These in their turn are the food of parasites and carnivores. While food is abundant and the weather is mild the animals of the coniferous

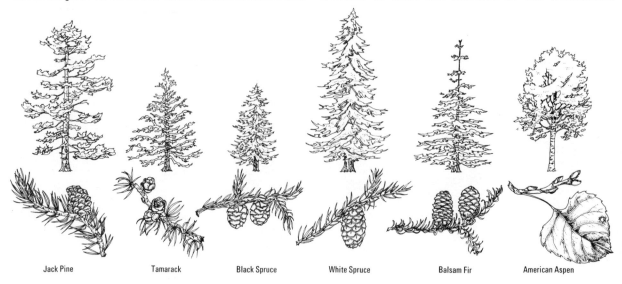

Jack Pine   Tamarack   Black Spruce   White Spruce   Balsam Fir   American Aspen

18

forest must produce and rear their young. For the birds this means that nesting must start in the spring, so that the young are hatched when the weather is warmest, and plenty of food is available. The larger mammals, such as the moose, whose unborn young take longer to mature, mate in the autumn so that the young are born in the spring. Only the bears produce their tiny young in the winter, having mated during the previous spring, and the cubs spend their first months with their mother in the warm security of the den, being fed only on milk. By being born during the winter they have the longest possible period in which to grow before having to face the next winter's harsh conditions.

With the coming of autumn the migrants leave the coniferous forest. The animals that remain feed hungrily, and some instinctively set aside food stores. These non-migrants hibernate, or seek sheltered dens during the winter, or they remain active, finding what food they can during the short bleak days or the long, bitterly cold nights. Winter provides the animals of the coniferous forests with their greatest test. Many of them fail to survive. It is the time when the weakest of the young animals die, and when the powers of the old falter. Both are likely to succumb to the icy conditions and, in so doing, to provide food for some of the luckier survivors.

When the spring comes the surviving animals will be the hardiest and fittest. The returning migrants, too, are severely tested by their journey. The parents that raise their young during the succeeding summer are the strongest and best adapted members of their species. In the very nature of things they are likely to have healthy and well-adapted young, capable of continuing the endless cycle in the years to come.

The biting cold and the short days of winter pose a daunting problem to the creatures of the northern forests—especially the small, cold-blooded ones. This is why many of the insects of

Using her saw-edged ovipositor—an egg-laying tube that in appearance resembles a sting—the female ichneumon wasp cuts a shaft through which to lay her eggs on the unresisting body of a wood wasp larva. When the young ichneumon larva hatches it feeds parasitically on its living cousin, and then spends the winter with the remains of its host, as shown in the life cycle diagram, emerging in early summer as an adult wasp. Since all forms of animals have their parasites, parasitism as a way of life is exceedingly common.

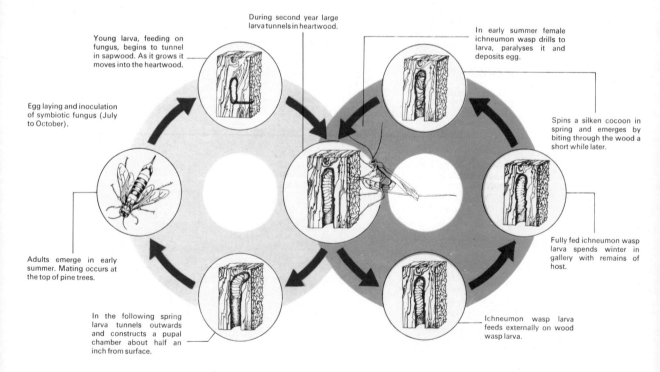

During second year large larva tunnels in heartwood.

Young larva, feeding on fungus, begins to tunnel in sapwood. As it grows it moves into the heartwood.

In early summer female ichneumon wasp drills to larva, paralyses it and deposits egg.

Egg laying and inoculation of symbiotic fungus (July to October).

Spins a silken cocoon in spring and emerges by biting through the wood a short while later.

Fully fed ichneumon wasp larva spends winter in gallery with remains of host.

Adults emerge in early summer. Mating occurs at the top of pine trees.

In the following spring larva tunnels outwards and constructs a pupal chamber about half an inch from surface.

Ichneumon wasp larva feeds externally on wood wasp larva.

Host: Wood wasp *(Urocerus)*
Two year life cycle

Parasite: Ichneumon wasp *(Rhyssa)*
One year life cycle

Like others of its kind, the North American pileated woodpecker drills out a nesting-hole in a solid tree trunk. It clings to the tree by means of its sharp, curved claws, bracing itself with its tail.

these forests have a long, inactive stage in their life history. The female sawfly, for example, mates in the early summer and then lays her eggs. At the end of her abdomen she has a short, tube-like egg-laying organ bearing tiny sharp saws with which she cuts a slit in a pine needle and lays an egg into it, repeating the process over and over again on other needles. Each egg hatches into a small grub or larva that feeds greedily on the pine needles. In late summer, when it is fully grown, the larva, too, faces the problem of the impending winter. It descends to the ground, buries itself, and spins a cocoon. Inside this it turns into a pupa and so passes the winter protected from extreme cold—as well as from most of its enemies. When warmer spring days return the winged adult insect hatches from the cocoon, ready to fly, mate, and start the life cycle again.

The giant wood wasp is closely related to the sawfly, but its egg-laying organ, or *ovipositor*, is adapted for boring into wood, and the female lays her eggs beneath the bark of pine trees, often choosing recently felled trunks. The larva that hatches from the egg bores farther into the wood, growing slowly and taking up to three years to mature. It feeds on wood that is permeated by a fungus. There is evidence that spores of the fungus are actually introduced by the female when she lays her eggs, thus ensuring that the grub will be provided with a suitable diet. These larvae do a great deal of damage to commercial timber.

The most effective natural enemy of the wood wasp larvae is a large ichneumon wasp called *Rhyssa*, whose body is over an inch long, and is armed with an ovipositor twice as long again. This organ is contained in a sheath furnished with sense organs that locate the wood wasp grub through an inch or more of wood, probably by detecting minute vibrations from its feeding and movements. With her slender ovipositor the female *Rhyssa* quickly bores down to the helpless victim, and, squeezing out an egg, deposits it on the grub, which is then doomed to a long, slow death. The ichneumon larva feeds on it, reaching maturity at the time when the unfortunate wood wasp larva dies.

Although many of its larvae perish in this way, and though it is quite defenseless and edible, the adult wood wasp is less liable to attack by birds than are most insects of its size. This is because it strongly resembles a fearsome hornet, which most birds take care to leave well alone.

The tiny goldcrest usually raises two broods, each consisting of between 7 and 12 young, during the short northern summer. The male helps to feed the young, but does not sit on the eggs.

The black-capped chickadee, another North American bird, eats insects in prodigious numbers, using its pointed beak to pick them from crevices and the surface of leaves.

Few insects of the coniferous forests can rely on such obliging "living larders," and most of them have to contend with resin if they feed on the most abundant food around—the conifers themselves. Consider, for instance, some caterpillars that feed on young pine needles. When resin oozes from the damaged plant the caterpillars instinctively seal it off by spinning a silken wall. If they did not, the sticky flood might ensnare them, and would in any case prevent them from continuing to feed at that point. Many of the beetles of the coniferous forests overcome this same difficulty by feeding only on dead and dying trees in which resin is no longer being produced. One of these beetles, the ambrosia beetle, has evolved an ingenious solution to the difficult problem of how to digest wood. Like the wood wasp, this beetle is always associated with a fungus, the spores of which are carried by the beetle in pockets on its body, and spread to any dead trees that it finds. The fungus is able to digest wood, and itself forms the beetle's food.

Crossbills make their untidy nests in trees at the edge of forest clearings. The adult's asymmetrical beak is well adapted both for shearing through twigs needed for the nest and for cutting the tough scales of pine cones to expose the edible seeds.

The spruce grouse of North America is the only member of a family in which the males have similar plumage to the females, and have only one mate. Unlike all other species of grouse, the male spruce grouse helps the female in the care of the young.

The insects of the coniferous forests are the main food of many kinds of birds but, as we have seen, it is only during the short summer that insects are at all abundant. Many species of birds raise their young in this brief time of plenty, then, when the days shorten and the temperature starts to fall, they migrate southward for the winter.

Prominent among the resident insect-eaters are several species of tits, chickadees, and nuthatches. In a way it is rather surprising that such small birds should be able to withstand the cold of northern winters, for small bodies are much less efficient at conserving heat than are larger ones. However, by trapping air, which is a poor conductor of heat, feathers provide superb insulation against the cold. When the temperature falls the resident birds ruffle their feathers and thus trap an even thicker blanket of air. During the short winter days the tits climb acrobatically among the trees searching every crevice for insects or their pupae. No branch remains un-

In the coniferous forests of North America, Clark's nutcracker uses its long, pointed bill to hammer pine cones to pieces in order to obtain the nourishing seeds that form a major part of its diet.

explored. So thorough is the search that it might seem that no insect could escape detection. Most do not, but insects lay so many eggs that the few survivors are enough to repopulate the forest during the following summer.

Of the other birds that survive the cold winters in the coniferous forests, woodpeckers are well adapted for finding food. Using their strong, pointed beaks as picks, they obtain insects that burrow into timber, and so are inaccessible to other birds. Woodpeckers have very long, flexible tongues with barbed tips by means of which they extract beetle grubs from galleries within the timber of a tree. They gain the leverage to do this by clinging to the bark with long, needle-sharp claws and using their stiff tail feathers as props. Other resident birds feed on the seeds of forest conifers.

When spring comes to the coniferous forests of the world the hardy resident birds are joined by waves of returning migrants, including several species of warblers. Although many species of warblers look much alike, each has its own distinctive song. Each pair of warblers establishes a territory that they defend against other members of their own species. Among the trees the rival pairs glimpse each other only occasionally, but their song, which is repeated almost constantly at this time, provides an unmistakable signal that the territory is occupied. The precise function of the territory is by no means clear. At one time it was thought to be primarily a food gathering area, but recent research suggests that certain species of warblers obtain most of their food while trespassing in neighboring territories. Perhaps territorial behavior serves mainly to space out the members of a species, preventing overcrowding.

By the end of the summer the migratory birds have raised their broods. Their young are fully fledged and able to fly. The days start to shorten, and this is the signal for migration. On arrival at their destinations the migrants feed on the leavings of the permanent residents, and may even change their diet for a season. The European brambling, for example, feeds on insects in the coniferous forests during the summer, but it spends the winter feeding on the fruits and seeds of the more southerly deciduous forests. Flexibility is often the key to survival.

The supply of insects as food in the coniferous forests varies greatly at different times of the year, but the supply of plant material remains much more constant. Despite their lack of teeth, some birds are able to feed upon the tough plant material that the coniferous forests provide. They can grind up leaves and seeds by means of the *gizzard*—a muscular part of the alimentary canal located just before the stomach.

The most striking of these herbivorous birds belong to the grouse family. They have feathers growing on their ankles, and during the winter often on their toes as well. These feathers may be of some use in helping to keep the bird's feet warm, but are more important as snowshoes, spreading the weight of the bird over as large an

The flying squirrel's "wings" consist of flaps of skin stretched between the front and hind limbs. They are almost invisible when the limbs are not extended, and in no way hamper the squirrel's busy tree life as it searches for nuts and fruits.

area of snow as possible. The nearly turkey-sized capercaillie of the coniferous forests of Europe and Asia is the largest of the grouse family. In spring the handsome, black-plumaged male displays and calls from a prominent branch, spreading out his fan of tail feathers as a courtship signal.

The capercaillie is polygamous, and a male may gather a harem of as many as a dozen mates at the same time. Since equal numbers of males and females are hatched, this means that many males have no mates at all, and during the breeding season the rivalry for females is intense. The females are smaller with brown feathers and usually nest on the ground where they hatch and rear their young unaided by the male. The dull coloring of the female provides good camouflage for, if she were as brightly colored as her mate, she would inevitably attract the attention of a prowling lynx, marten, or bear. The young can run and feed themselves as soon as they are hatched. Their diet consists mainly of buds and shoots of conifers although, like other game birds, they also scratch in the ground for seeds, insects, and other invertebrates.

The seeds of the coniferous forests form the food of several species of small birds belonging to the finch family. The crossbills, for example, have asymmetrical bills with overlapping tips that they use for scissoring through the tough scales of unripe cones. They can then extract the exposed seeds with their long tongues. Crossbills are the only finches that may spend the whole year in the coniferous forests, migrating southward only during the most severe of winters. Twites and linnets, another group of old world finches, spend the summer on the Arctic tundras, but in fall feed on tree seeds as they migrate southward through the forest belt.

Squirrels are particularly successful in tapping the supply of plant material in the coniferous forests as a source of food. Their sharp incisor teeth can shred bark and leaves, which are afterward passed to grinding teeth toward the back of the mouth to be thoroughly chewed. For the agile forest squirrels trees are a source of food and a means of escape form heavier enemies.

Apart from their small size climbing squirrels have another obvious adaptation for climbing; they have longer tails than nonclimbing squirrels. This extra length aids in balancing. Generally, however, there is a notable absence of features normally associated with climbing animals.

For example, squirrels' eyes are toward the sides of their heads and this should mean that squirrels cannot judge distances very accurately, an ability essential when jumping from branch to branch. In addition, their fingers and toes are not very long, and can grasp only the smallest of branches. However, what they lack in adaptation they more than make up for in sheer agility. Although squirrels often pause when they have a secure foothold, they nearly always take the most difficult "treeways" at great speed.

The seeds of trees form an important part of the squirrels' food. Seeds are most abundant in the early autumn, and at this time squirrels store some of the surplus, putting them into any convenient tree-hole, and even burying them in the ground. This habit of storing food is purely instinctive, and ensures that some food will be available during the winter, when squirrels are less active but do not hibernate, and therefore

The North American red squirrel is less brightly colored than its European cousin. Perhaps because it has to face even more extreme conditions, it makes even larger winter food stores, hiding seeds and nuts away in tree-holes and burying them in the ground.

need a regular food supply. Since squirrels re-locate their stores by accident, some seeds may be easily overlooked and this helps the spread of trees.

When leaping from one slender branch to another, squirrels spread their limbs and their long bushy tails as they glide through the air. It is not altogether surprising that some members of the family have evolved folds of skin, stretched between the fore and hind limbs, which help them glide from one tree to another. These flying squirrels are only slightly larger than mice, and they can glide for distances of up to 50 yards,

inevitably losing some height on the way.

Clambering slowly through moonlit branches, the nocturnal Canadian porcupine provides a stark contrast to the agile, daytime leaping of the squirrels. The porcupine has no need of speed as a defense against enemies, for hidden in its warm fur it has sharp, barbed quills. It spends the daylight hours in a den, which may be either under the ground, or among the branches of a tree. Canadian porcupines do not hibernate or store food. The tree bark that forms the major part of their diet is available the whole year around. None of the other larger herbivores wanders

Right: during the summer moose often wade in shallow lakes in order to feed on waterweed. The water also provides some relief from the summer heat, which can be considerable. At this time of the year the antlers of bull moose have completed their growth, and the furry skin, known as velvet, that has covered them during the spring dries and peels away.

Chipmunks are burrowing members of the squirrel family. They sometimes feed on the seeds and buds of coniferous trees, but can be found mainly feeding on the fruits and seeds of deciduous trees that thrive at the edges of clearings in the coniferous forests.

around among the branches with such freedom; they live firmly on the ground.

Moving southward from the bare Arctic tundras to the shelter of the coniferous forests for the winter, caribou and reindeer may join other deer, such as moose, that spend the whole year in the forests and also, occasionally, species of deer more typical of the deciduous forests to the south.

Moose usually live in marshy coniferous forests, where they browse on willow, or wade and swim in the shallow lakes for a meal of waterweed. In winter, when the willow has lost its leaves and the lakes have frozen over, they feed on the shoots and bark of conifers. Moose are solitary animals, and each individual usually keeps to one small area of forest, which it does not defend—indeed there are times when it may share this with other moose.

In early autumn, the bulls fight over the females, clashing antler against antler and bellowing furiously. A bull does not gather together a group of females, but may mate with one female after another during the month or so of the rutting season. The young are born in May or June of the following year. They can run almost immediately and they grow rapidly, reach-

ing adult size within three years. The unusually broad hoofs of the adult moose spread out the animals' weight so that they can move easily over soft ground. Only when there is deep, soft snow do moose find difficulty in moving through the forests. During this hard time they often gather in large herds and stamp "yards" in the snow. These trampled areas uncover berries and small shoots that are just sufficient to keep the animals alive until spring. It is under these conditions of restricted movement and near-starvation that moose are most likely to fall prey to a bear or to a pack of wolves.

Completely hidden from view beneath the surface of the ground, voles and wood lemmings find protection from their natural enemies and from the worst of the weather. Being small and squat they can burrow with ease. They remain active even during the long, hard winter, making small tunnels in the snow through which they can move unseen in their busy search for food.

Chipmunks, too, make their homes underground, emerging to climb among the lower branches where they feed on leaves and fruits, and occasionally upon birds' eggs and insects. Their winter sleep is only shallow, and at intervals they wake up to feed. The more heavily built woodchuck sleeps more soundly in its grass-lined burrow, hibernating for up to eight months of the year. Only during the warmest months do woodchucks emerge to feed on grasses and insects.

Beavers, by contrast, are certainly the busiest mammals of the coniferous forests. Working together, a pair or a family of beavers constructs a lodge of mud and sticks. The entrances are beneath the water of a lake or river, but the snug den inside the lodge must be above water level. This is not difficult to achieve in a lake, but in running water beavers must construct a dam, the

Like mankind, beavers shape the landscape to their own short-term advantage, damming streams in order to form the shallow lakes that they love. American beavers build a dam of branches, stones, and mud, and make their home in the shallow lake thus created. Their lodges are usually about 6 feet in diameter, with underwater entrances. Fresh branches are piled in the water nearby as a food-store. Each lodge is occupied by a single pair of beavers and their young. Young beavers are born within the lodge at the beginning of the summer. At this time the male leads a bachelor existence in a burrow at the edge of the beaver pond.

longest of which may reach several hundred yards and contain hundreds of tons of mud and branches. All of it must be laboriously hauled into place. In addition, beavers work hard during the summer months to amass a winter food store. They relish the bark of deciduous waterside trees, particularly aspen, birch, and willow. With their chisel-shaped front teeth they fell such trees in order to reach the tender inner bark of the topmost branches. They gnaw off these branches, drag them into the water, and store them in a stack near one of the entrances to the lodge in preparation for the long winter.

Beavers mate during the winter, and the young are usually born in April or May. At first they are helpless, and feed only on the mother's milk, but by the time they are two months old they start to eat leaves and young, soft shoots. Young beavers grow slowly and do not attain full adult size— about three feet long from nose to tail-tip—until they are over two years old. Until they reach this age they remain with their parents.

Beavers are superb swimmers for their webbed feet make powerful paddles. Their fur is so thick that, despite prolonged immersion in water, some air always remains trapped next to the beaver's skin, insulating it against the cold even when the beaver is swimming beneath the ice of a frozen lake. When danger threatens, beavers can remain submerged for up to 15 minutes. Under the water they are relatively safe from their enemies, but

Above: the goshawk, one of the fiercest birds of prey of the coniferous forests, is widely distributed in North America, Europe, and Asia. It catches its prey by speed combined with maneuverability, twisting and turning among the treetops. It hunts birds, squirrels, and sometimes martens.

Right: an American saw-whet owl swoops on the prey it has located, using its large eyes and its keen sense of hearing. The prey has no warning of impending danger, for the owl's soft wing-feathers make no sound as they move through the air.

32

when they venture on to land for food or building materials they are more likely to fall victim to wolves, bears, or lynxes.

The lakes that form upstream of beavers' dams cause considerable changes in the environment. In the course of time a beaver pond may become silted up and turn into a grassy open space that eventually becomes tree-covered. If a new generation of beavers sets up a colony by damming the original stream at this time, the whole cycle may repeat itself as a new lake is formed.

For the smaller animals of the coniferous forests birds of prey are a constant menace. The larger ones such as the golden eagle, penetrate no deeper than the forest edges, because they cannot maneuver well enough to chase their prey among the trees. Where the trees grow more thickly the goshawk is the most powerful bird of prey.

During the day it is the turn of the hawks and buzzards. While daylight lasts their eyes work superbly, but when evening comes the hawks

must roost until morning because they cannot see at all in dim light. For the hunted animals of the coniferous forest this does not mean safety, for at night the owls take over. Owls can see by the light of the stars on a moonless night. Their specially adapted eyes respond to light of very low intensity, and although they are not able to distinguish fine detail or color, they can see better at night than any other animal, and are certainly able to see well enough to hunt.

Predatory birds of both the night and the day have very strong toes armed with powerful talons, and these are the weapons with which they usually seize their prey. Such hawks as the goshawk and the merlin rely on speed when hunting, while the American sharp-shinned hawk and the European sparrowhawk depend more on maneuverability when they pursue small birds among the trees. Owls rely on their ability to fly noiselessly on soft-feathered wings so as to surprise their prey. The more powerful predatory birds hunt smaller birds and such mammals as squirrels and voles.

Predatory birds can never be as numerous as the animals they hunt. Ideally they should leave enough of the hunted species to breed and maintain the population. However, this balance between hunters and hunted is not always achieved in the coniferous forests. The population of small rodents, such as voles and wood lemmings, instead of remaining roughly constant from one year to another, tends to vary. The numbers build up over a period of years until a peak is reached. Then, because of disease, or food shortages, or perhaps even nervous exhaustion due to overcrowding—the precise reason is not known —large numbers of rodents die, and the numbers fall to a low level, from which they gradually increase once again. Predatory birds, such as the long-eared owl, which feed primarily on these rodents, show fluctuations in numbers geared to those of their prey.

On the ground, too, there are deadly hunters ready to pounce upon the unwary. The northern lynx, for example, stalks its prey until it can pounce from close range. A solitary hunter in the coniferous forests, the lynx has large feet that

If a full-grown moose is brought to bay by a pack of wolves and stands its ground, it has a good chance of survival. It is the moose that keep running and do not turn and face their attackers that are most likely to fall prey to the wolves.

enable it to move over soft snow with ease and hunt stealthily for small deer, and large birds such as grouse.

Wolves, too, are greatly feared by the deer of the forests, perhaps even more than the lynx, for several of them hunt together. There are usually up to a dozen wolves in a pack, which consists of an adult pair and their last one or two litters. Although they cover long distances when hunting, they usually travel in a circle and thus remain within the same area.

Small rodents and even some vegetable food such as fruits form part of the wolves' diet, but more usually wolves hunt larger animals such as reindeer and moose. When a suitable victim is located the wolves chase after it, loping along easily at up to 28 miles an hour. If they lose sight of their quarry the wolves spread out and, using their keen noses, search for the scent. When they find it they resume the chase. They are long-distance runners and if necessary may keep up the pursuit for 10 miles or more. When they do catch up with the exhausted prey they surround it, and soon overcome it.

Despite their speed and ferocity, wolves are not always successful in their hunt. Young, inexperienced deer and injured or aging ones are their most likely victims. Because of this, the effect of predators such as wolves on the populations of animals that they hunt is in many respects beneficial. By killing off the weak and careless, predators ensure that the surviving members of the prey species will be healthy and wary, and thus the best possible breeding stock.

In the coniferous forests wolves usually mate during February or March. Four to six cubs are

Mink usually live near water, where they rival otters in their swimming ability. They feed on fish and small aquatic mammals but may also take birds, as in the picture.

The short-tailed weasel is widely distributed in North America, and also in Asia and Europe, where it is known as the stoat or ermine.

born two months later in a den often located under the roots of a fallen tree. While the young are helpless the male hunts on his own, returning to the den with food for his mate. The cubs are weaned at about six weeks and when they are about three months old they leave the den, accompanying their parents in their wandering life. In the autumn the pack may be joined by the young of the previous year, which have had to fend for themselves during the summer. When spring comes again the two-year-olds will be driven away by the mother. They are now fully adult and old enough to breed and to found packs of their own.

Smaller than the wolves, but equally fierce among the predators in the coniferous forests are the members of the weasel family. With their long, muscular bodies and short legs, they are very adaptable, and hunt small mammals and birds in a variety of ways. On the ground stoats and weasels hunt for ground-living rodents, often entering the burrows of their prey, while the more heavily built wolverine, or glutton, preys on large birds, such as grouse, large rodents and, occasionally, deer. Overhead in the branches, martens and sables stalk birds and chase squirrels, for they have long tails and can climb well.

Constantly at work searching for insects, the shrews of the forest floor pack several cycles of activity and rest into each day in order to get enough food. Like the voles and lemmings, they do not hibernate, remaining active throughout the year and in winter making tunnels beneath the snow on the surface of the ground.

In their dens among rocks, in caves, or sometimes under the ground, the bears of the coni-

Being longer-tailed than other members of the weasel family, the marten is a skillful and speedy climber, able to catch and overpower squirrels.

ferous forests spend much of the winter sound asleep, but their sleep is not true hibernation, for their body temperature remains high, and because they use energy in keeping their bodies warm, they sometimes become hungry and emerge to feed during the winter. When warm-blooded animals truly hibernate their body temperature falls to only just above that of their surroundings. In this state of suspended animation they need very little energy.

Emerging with the coming of spring, bears have no fear of starvation, for they are omnivorous feeders. With their great strength, fearsome teeth, and powerful claws brown bears, for instance, can overcome any prey they can catch, no matter how large. But although they can move at 30 miles an hour, they cannot run far at this speed. Moreover, they are too heavily built to be very stealthy and, since they move on flat feet, are not well adapted for springing or pouncing, so it is not surprising that they are often unable to catch the largest animals of the coniferous forests, such as the moose, and instead have to dig for rodents, and raid birds' nests. These do not entirely satisfy the bears' hunger, so they also eat plant materials such as roots, leaves, and fruits. In the autumn bears feed hungrily and grow fat. During the cold winter this fat acts both as an insulating material to keep them warm, and as a built-in energy store.

Although its ancestors were exclusively carnivorous, the brown bear will eat any kind of food. For North American brown bears the time of plenty comes when the salmon, rich with eggs, swim up-river to their spawning grounds.

A Simple Food Web in a Coniferous Forest

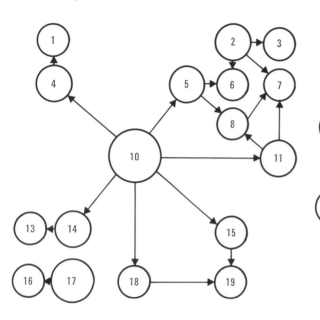

1 Pine cone
2 Red squirrel
3 Balsam fir cone
4 Red crossbill
5 Bay-breasted warbler
6 Spruce sawfly larva
7 Spruce budworm moth
8 Braconid wasp (parasite)
9 Aspen shoots
10 Goshawk

11 Golden-crowned kinglet
12 Beaver
13 Bark beetle
14 Three-toed woodpecker
15 Woodmouse
16 Wood wasp
17 Ichneumon wasp
   (parasite)
18 Common shrew
19 Sawfly pupa

The illustration shows a much-simplified food web for a North American coniferous forest but could also apply, with very few changes, to the northern coniferous forests of Eurasia. The dominant trees in these forests are the evergreen, needle-leaved pines and firs, such as the eastern white pine and balsam fir illustrated here. The clearings and marshy areas, however, are dominated by such broadleaved deciduous trees as the willows, aspens, and birches. These are important as major food sources for such characteristic plant-feeders as the beaver and moose, as well as innumerable kinds of insects. Away from these clearings little light filters through the closed canopy, which accounts for the general absence of shrubs and herbaceous plants. This lack of structural diversity, together with the harsh seasonality, is largely responsible for the relatively small number of animal species in the forests.

Coniferous forests are also often characterized by remarkably even-aged stands of trees caused by natural and man-made fires. Fire spreads rapidly through the dry needle litter, which is prevented from rapid decomposition by the cold winters and the inhibiting effect of resins and other compounds. In the case of the jack pine in North America, fire is actually needed before the seeds can be liberated from the cones. In such uniform conditions only a few species tend to become dominant—sometimes becoming extremely abundant as shown by the spruce budworm, which occasionally defoliates large areas of forest.

Temperate Forests

The world's temperate forests grow only in moderate climates where the rainfall is plentiful. For this reason they do not, like the coniferous forests, form a continuous belt across the Northern Hemisphere. The forests are broken up by great mountain ranges such as the Rockies, which cast a dry rain-shadow on the side away from the prevailing winds, and in any case the parched heartlands of the great continents are far from the sea and beyond the reach of rain-laden winds. They are too dry to support forests.

The cooler temperate forests are chiefly made up of deciduous trees, such as maple, hickory, oak, and beech. During the warm summer, when sunlight is plentiful, these trees spread their broad leaves, and make the most of the favorable conditions for building new tissues. When summer ends a layer of cork forms across the base of each leaf stalk. Deprived of water the leaves wither and fall. All winter long the trees stand bare, their new leaves protected from the frost within tightly packed, dormant buds.

In slightly warmer climates, where the winter is less severe, evergreen forests grow. The trees of these forests, unlike coniferous evergreens, are broadleaved. The leaves have tough, protective cuticles similar to those of the needles of the conifers. Some of these broadleaved evergreen trees grow alongside deciduous trees in the cool temperate forests, but most are found in the southern United States, on the Mediterranean coasts of Europe and Africa, and in southern China and Japan. In Australia and New Zealand broadleaved evergreen forests entirely replace deciduous ones.

The temperate forests once covered almost the whole of the eastern United States and Europe south of Scandinavia, but the activities of man over thousands of years have destroyed most of them. Only an occasional wood or copse now remains. However, the plants and animals of these isolated clumps are not the only survivors of the ancient forests. In country hedges, suburban

Vividly caught by the camera, the seasonal changes of the trees of the deciduous forests, from full leaf to bare branches, impose a marked and regular rhythm on the forest's inhabitants.

Spring ▲

▼ Winter   Summer ▲   ▼ Fall

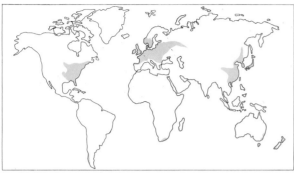

The World's Temperate Forests

The map at left shows the areas of the world's temperate deciduous forests. At the foot of this page are drawings of trees typical of the forests of North America, to the same scale as similar drawings on pages 18 and 82. Below the trees are details of their foliage. On the facing page sunlight floods across the floor of a deciduous forest. Compared with the coniferous forest shown on page 19, the deciduous forest has more undergrowth, a more varied canopy layer, and more species of animals.

gardens, and city parks the plants and animals known to millions of people are really those of the temperate forests. The same pattern has occurred in much of China, Japan, the nontropical coastal fringes of Australia, and in New Zealand.

Deciduous trees follow a steady seasonal rhythm of leaf fall but the leaves of evergreen trees age, wear out, fall, and are replaced at all seasons of the year. In both cases the result is the same: the forest floor becomes thickly carpeted with dead leaves. At first sight this looks like sheer extravagance. Minerals obtained from the soil, forming part of plant materials built up over the months appear to be cast away and wasted. However, in nature there is no real waste. These fallen leaves contain no preservative resin, so bacteria, fungi, and tiny animals can quickly break them down, causing them to decay. The minerals are returned to the soil, and from there can once again be absorbed by the plants and used in the manufacture of new plant material.

Nowhere on earth are the changing seasons more marked than they are in the deciduous temperate forests. The contrasts between the fresh promise of spring, and the mellow fruitfulness of the fall could hardly be greater, and between these seasons come the fulfillment of summer and the barrenness of winter.

In order to survive in such a changing environment the animals of the deciduous forests must be adapted either to withstand or to avoid the testing time of winter. For the most active animals, especially the birds, migration is sometimes the best answer, because activity requires food, and this is often scarce in the winter. At the other end of the activity scale, the cold-blooded vertebrates migrate only locally or not at all. The falling autumn temperature chills their bodies, slowing down their life processes so that for them hibernation is an inevitable response to winter.

Some of the warm-blooded animals of the deciduous forests also hibernate during the winter. For them, this means that the automatic control

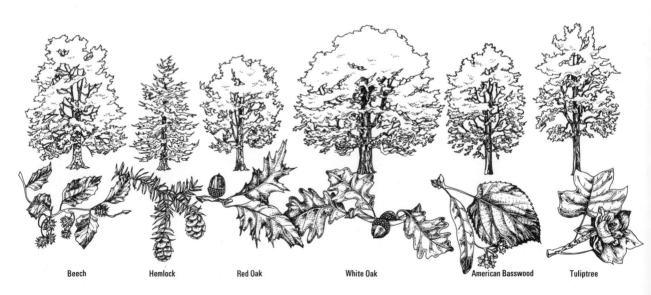

Beech   Hemlock   Red Oak   White Oak   American Basswood   Tuliptree

of body temperature that is their characteristic feature is abandoned, so they do not use up food merely to keep warm. Other warm-blooded animals remain active throughout winter, perhaps sleeping more than in the summer, but maintaining a high body temperature while doing so. These animals must have a constant food supply, and it is to meet this need that some herbivores have evolved the habit of storing food instinctively during the glut that the fall brings. Migration, hibernation, or continuous activity are different solutions to the same problem.

In broadleaved evergreen temperate forests nearer to the tropics, conditions are less extreme, both for the trees that make up the forest, and for the animals that live there. However, even here there are seasonal changes. For example, flowering and fruiting take place at fixed seasons, and the availability of pollen, nectar, fruits, and seeds as food for animals varies accordingly. Even in these forests, winters can be cool. Nevertheless, these broadleaved forests provide a less harsh environment for animals than do the deciduous forests farther north. Conditions are closer to those of the lush, humid forests of the tropics, with which the evergreen temperate forests sometimes merge and blend.

The constant decay of leaves gives temperate deciduous forests a rich, dark soil whose surface layers support a teeming wealth of invertebrates. Earthworms feed on the decaying materials in the soil and, in doing so, act as natural cultivators. Over the years they bring buried materials to the surface in the form of worm-casts, and their burrows aerate the soil. They turn over the ground and break it up as effectively as any farm plow. Smaller, but equally important, are the myriads of tiny, wingless insects, such as springtails, which feed on the decaying leaves or on the fungi that promote decay.

Here and there in the temperate forests lies a fallen branch or an entire tree—an ancient giant brought low by the insidious activity of fungi within its trunk, and perhaps finally toppled by wind or lightning. Crevices develop within the rotten heartwood of its trunk and under its gnarled, peeling bark. Here, and on the ground underneath, can be found many small herbivorous animals—snails and slugs, arthropods such as woodlice, teeming insects, and millipedes. The crevices provide a degree of safety from the constantly searching, ever-hungry carnivores.

Being sheltered from the wind the crevices are

Above: like the larger predator after which it is named, the wolf spider catches its prey by running. Unlike the wolf, however, the wolf spider has little stamina, and must stop every few yards in order to rest, before continuing the chase.

Left: The pseudoscorpion is a relative of the spider. Although tiny (they are less than a quarter of an inch long), pseudoscorpions are very fierce, killing their prey and tearing it apart by means of the pincers, which are armed with poison.

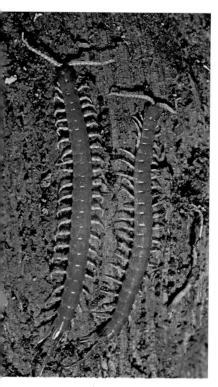  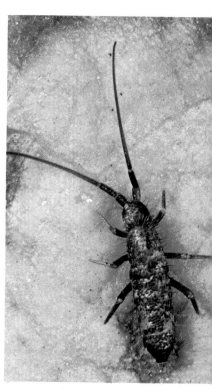

The temperate forests harbor a wealth of invertebrate animals. Left: centipedes are found in the soil and in damp places under leaf litter, stones, or logs. They are carnivorous and kill their prey with poison fangs. Center: the woodlouse is herbivorous, and hides out in moist places, often under rotting logs. Right: the herbivorous springtail lives in the surface layers of the soil, feeding on decaying leaves. Springtails are so-called because they leap with a forked organ, which is folded under the body when the animal is resting.

There are two pairs of legs on each segment of the millipede's body, but despite this it cannot run as fast as a centipede, which has only a single pair of legs on each segment. Millipedes use their legs for clinging as they climb in search of food. They are herbivorous.

usually rather damp and this, too, is important for many animals. Some arthropods, such as the woodlice, breathe by means of special gills, which must remain damp if they are to work. Snails and slugs may venture out to feed, especially after rain or at night when the temperature falls and the air yields its moisture as dew, but without the perpetually damp crevices to hole up in they could not survive—particularly the slugs, which lack a shell that could help prevent water loss.

Small carnivores frequent the same dark haunts as their prey. Obviously it is convenient for them to be near their food, but that is not the only reason. Small carnivores are preyed upon by larger ones, so the crevices provide a welcome degree of safety for them, too. Another reason is that some of the arthropod predators—many spiders, for instance—breathe partly by means of damp, external gillbooks, and so need a moist environment. The predators include a great variety of spiders that are mostly small and inconspicuous, for, contrary to popular belief, only a minority build prominent and exposed webs. There are also centipedes, and tiny pseudoscorpions, which look like real scorpions, but with venomous fangs on their claws instead of a sting at the tail.

Feeding upon all of these invertebrates are vertebrates, small from our point of view, but efficient and deadly to their tiny prey. Shrews, for

Above: some gall wasps lay their eggs on oak leaves, which form spangle galls around the developing grubs. When the galls mature they are shed from the leaf, and the larvae inside continue to develop on the ground. The grubs overwinter in the galls and adult gall wasps finally emerge in the spring.

Mantises are often protectively colored to blend with their surroundings. They lie in wait for smaller insects and then seize the prey with their powerful forelegs. The picture shows a bark-mimicking mantis eating a smaller green leaf mantis. The sharp spines on the legs impale the insect and prevent it escaping.

instance, are even more abundant than they are in the coniferous forests, and every few hours of the day and night, a shrew must wake, scour the forest floor for food, eat, and return to rest.

Shielded among the fallen trees and the tree roots, and hidden away in the soil or crevices, the invertebrates of the temperate forest floor are less subject to seasonal changes than those that live among the branches of the trees. Especially in the deciduous forests, the winter is a difficult time for tree-living insects. They often depend upon the leaves for food, and always need them both as shelter from the worst force of the wind and rain, and as a shield against the searching eyes of predators. When the leaves fall, food, shelter, and safety are all lost, so the life cycles of these tree-living invertebrates closely resemble those of the invertebrates of the coniferous forest. They over-winter, some of them as eggs, and, among the insects, frequently as larvae or pupae.

Even when the trees are in leaf these animals are more exposed than those of the forest floor, but like most living things they are well adapted to their environment and are often well camouflaged. Both herbivores, such as leaf-eating caterpillars, and carnivores, such as mantises, are frequently green, colored by pigments known as bilins, derived from the green plant material, chlorophyll. Camouflage does not depend on color alone, but sometimes on shape, and always on behavior. Different species of mantis, for example, have leaflike, or twiglike bodies. During daylight when the keen eyes of insect-eating birds function superbly, hawk-moth caterpillars keep still, their green bodies mimicking the leaves, and the mimicry is astonishingly good. They have markings like the veins of leaves on their sides, and a long thin projection, which looks like a leaf stalk, near the tail. Some moths have patterns on their wings to blend perfectly with the lichens that grow on tree-trunks, and even gaily colored butterflies can virtually disappear when they fold their wings together, concealing their colors and exposing only their wings' dull and camouflaged undersurfaces.

Some insect larvae burrow into the trees for

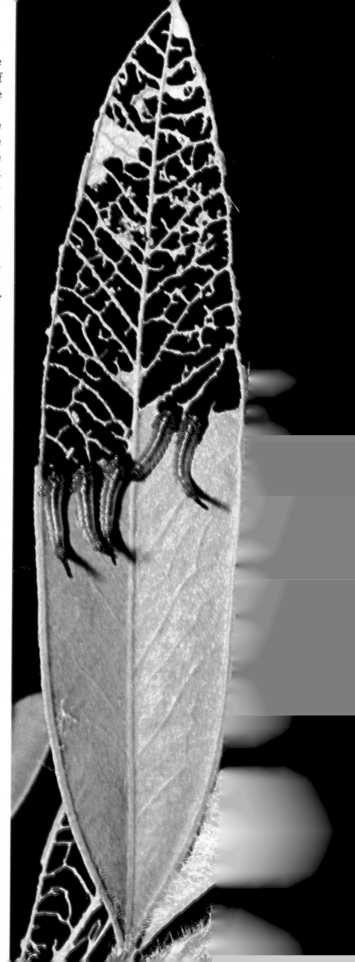

Hungry caterpillars feed side by side, stripping these leaves to a lattice of veins. Leaf-eating caterpillars form an important part of many forest food chains as prey for birds and other insect-eaters.

During the early part of the breeding season the long springtime chorus that gives this tree frog its name of "spring peeper" is a familiar sound to many people in the eastern United States. Despite their noisy song, spring peepers are easily overlooked because they are small—only about one inch long—and their skin color can change to match various backgrounds.

This wood frog is a common and attractive resident of the northern wooded areas of North America. Unlike many other frogs it is not restricted to a waterside life and wanders freely, returning to water only during the breeding season.

With the vocal sac distended in song this male gray tree frog clings to a creeper, well camouflaged against the tree bark. Like all tree frogs it is especially adapted to life in the trees, with suction pads on its toes that enable it to grip branches and leaves.

both safety and food. The many different kinds of leaf-miner that feed inside leaves, constructing contorted galleries, are the larvae of flies, saw-flies, moths, and beetles. Here is a case where the members of several groups of insects have, quite separately, evolved the same adaptive mode of life, a process known as *convergent evolution*.

The spittle bugs illustrate yet another way in which insect larvae are sometimes protected. The larvae develop on leaves inside a mass of froth, which they create for themselves by blowing bubbles from the last few segments of the abdomen. This froth is distasteful to many predators.

Some of the safest insects are those protected within galls. Some trees, when they are damaged by the presence of an insect, respond by forming growths of tissue, which seal off the invader, and minimize the damage caused. These growths are called galls. They vary in size from the width of a pinhead to more than an inch in diameter. Some are compact and look rather like fruits, for example, oak apples. Others have a more branched appearance. The plant tissues forming the gall provide both food and protection for the insect inside. Most galls are caused and used by the larvae of tiny gall wasps, but some are caused by gall midges, and others by aphids.

Aphids damage trees and plants not only by causing galls to form but also by sucking the rich, sugary sap. It may seem strange to think of the inconspicuous little aphid as such a menace, but they breed so rapidly during the summer that their sheer numbers—all drinking sap—can dangerously weaken trees and shrubs. The numbers of aphids are, however, kept in check to some extent by other insects such as the larvae of ladybug beetles and lacewing flies, which feed on them.

Unlike the invertebrates, many vertebrate animals are relatively large and need very efficient limbs for movement on land—so much so that on a really cold day only a warm-blooded animal can achieve the necessary energy output. For this reason cold-blooded land vertebrates— amphibians and reptiles—occur only rarely in coniferous forests, and even then only at the southern fringes. In more temperate regions the summer sun is strong enough to raise the body temperature of a cold-blooded vertebrate to a reasonable level and amphibians and reptiles certainly occur in these regions even though they must hibernate each winter. Their way of living does, in fact, give them certain advantages.

51

Because they gain body heat from their environment, instead of using their food as a source of heat energy, they can go for a long time between meals, weeks if necessary. If they cannot move with great speed, at least they are good at keeping still, and this is a great aid in hiding, either from enemies or when ambushing prey.

Amphibians depend on water for breeding, but not all of them need a large volume of standing water. The widespread woodland salamanders of North America, for example, lay their eggs in small groups among damp, rotting logs or moss. They have no water-living tadpole stage. Instead a small, fully developed salamander emerges direct from the egg and feeds at once on insects and other invertebrates. Other North American amphibians breed more conventionally, and most are never found very far from ponds or lakes.

Ground-living amphibians like the American toad, and smaller climbers like the spring peeper, a tree frog with adhesive disks on its toes capable of clinging to twigs or leaves, breed in water. Like other tree frogs, the spring peeper can change its camouflage to match various backgrounds.

Lizards that climb well, such as the fence lizard and its numerous relatives, not only hunt insects and small invertebrates in the trees, but also climb the nearest tree for safety if they are disturbed when on the ground. When a fence lizard or a tree lizard freezes against a branch its grayish body patterned with darker markings is well camouflaged. The race-runner, a long-tailed lizard that often inhabits clearings, cannot climb, but springs to the nearest ground cover with remarkable speed when startled. Also sometimes living in open woodland, but rarely seen because it is an excellent burrower that spends much of its time below ground, is the legless and shiny-skinned glass-lizard. These lizards are easily mistaken for snakes, but can be distinguished because, unlike snakes, they have external ear openings and movable eyelids.

From forest floor to the open sky above the trees there are a great number of ecological niches to fill and, since any given ecological niche may be filled by different species in different areas, the number of species to be found in the broad span of the temperate forests is very large. Among birds alone there are fruit-eaters, seed-eaters, omnivorous birds, insect-eaters, and a variety of large predators. In North America, for example, barn swallows wheel and turn over the treetops in summer in pursuit of flying insects, while hummingbirds exhibit a life style unknown among the birds of the temperate forests of Europe and Asia by hovering and feeding on the nectar of summer flowers.

No part of the day or night is free from activity of some kind. For instance, the owls of the forest are usually thought of as night birds, but in America the barred owl and the great horned owl may be active during daylight, hunting for beetles, mice, or rabbits. At dusk the great gray owl wakes and joins the hunt, while by night the screech owl and the saw-whet owl take over.

As a group, the owls of the forest appear to occupy a relatively safe niche—one that is broad enough to support many species—but other forest niches are not as secure as this and species adapted to them are inevitably at greater risk. For instance, the ivory-billed woodpecker of North America feeds only on the grubs of flat-headed wood-boring beetles, and these insects thrive only in dying and newly dead trees. Since it takes a great many beetle grubs to enable just one pair of ivory-billed woodpeckers to raise their brood, the ideal habitat for these woodpeckers is an area with many dead and dying trees. Such areas have never been common in healthy forests and the ivory-billed woodpecker was therefore never a common bird. The destruction of the luxuriant forests of the southern United States by man may have been the last straw for this species. The ivory-billed woodpecker has not been seen for some years now. Perhaps it is extinct. The niche was, in fact, so narrow that the slightest pressure closed it.

By contrast, a forest animal that can thrive on, say, a wide variety of grasses, tree leaves, and young shoots, occupies a broader and much safer niche. Forest deer are a good example.

The white-tailed deer of North America's temperate forests usually live in small groups of three or four. Only when there is deep snow and food is scarce do larger herds collect in places where food, such as leaves or twigs, is most easily found. The adult's winter coat is thick, and grayish-brown; the summer coat is reddish-brown. This change of color with the seasons helps to camouflage the deer at all times of the year.

Like the adults, the fawns are well camouflaged. They are born in early summer, about seven months after their parents mated. At first the mother leaves them in the thickest cover that she can find, returning only to suckle them.

Northern waterthrush

A pair of ring-necked pheasants

A pair of cardinals—the male's plumage is bright red

The temperate forests of North America are home to a vast array of birds, some of which are shown here. The cardinal, the "red bird" of the southern states, belongs to the largest family of North American birds, the sparrows and finches. Members of this family have the short, conical bill characteristic of seed-eaters. They are found in a wide variety of habitats, both forested and open. The northern water thrush, despite its name, is a wood warbler. It lives and feeds near water, searching the ground for insects. Ring-necked pheasants are not native but have been introduced into North America and now live wild in open woodland and farmland.

Among the bushes the fawns remain still and safe, for their brown fur with white spots blends beautifully with the dappling of sunlight filtered through the leaves. After the first few days they follow the mother, and by this time they are already taking some solid food. They are weaned after six weeks, although they stay close to the mother until they are up to two years old, by which time they are fully mature, and have lost their spots.

The white-tailed deer owes its name to the white underside of its tail, and the white patch on its rump. When the deer is startled its tail stands up, exposing both its own white surface and that of the rump. To other deer this white patch is a signal that says, in effect, "Follow me." If only one deer within a group senses danger and runs, the others are automatically led to safety. The deer was an important source of food and buckskin for the forest-living Indian tribes. Deer still

exist in large numbers in North America and are a favored target for sportsmen.

The natural enemies of these forest deer once included a number of carnivores that have disappeared with the march of human civilization. Wolves, brown bears, and lynxes of the species that survive in the coniferous forests have vanished from much of their former range.

The smaller and perhaps more cunning carnivores still survive in some of the temperate forests. Once again these include some of those that also inhabit coniferous forests, such as both ground-living and climbing members of the weasel family. When we think of the way in which some kinds of animals are able to survive only in very specialized ecological niches, and must have just the right sort of climate, vegetation, and so on, it is surprising what a wide range of habitats other kinds of animals can tolerate. Some species can thrive in both coni-

53

ferous and temperate forests, and others can live not only in both types of forest but also in more open country. For instance, the American badger, a large and heavily-built member of the weasel family, is sometimes found at the edge of the forests of western North America, although it more often lives in dry, open plains. The striped skunk is even more tolerant, for it is at home in forests, plains, or even deserts, being equally adaptable in habits and diet, feeding omnivorously on fruit, insects, and small vertebrates.

Skunks range as far north as southern Canada. There they sleep soundly through most of the winter in underground dens, although they do not truly hibernate. Skunks usually stay out of sight during the day, and venture out at night in search of food. Their methods of self-defense are notorious. The bold black and white pattern of the fur is a warning, and to heighten its effect a startled skunk postures in a manner that emphasizes the coloring. If this warning is not heeded the skunk turns its back on its enemy and, looking over its shoulder in order to aim, it discharges streams of colorless liquid from two glands, one on either side of its rectum. These jets have a range of up to three or four yards, and as they pass through the air they vaporize to give not merely a truly appalling smell, but a poison gas that seriously interferes with the victim's breathing. The smell lingers for days— just to reinforce the message. On humans and their clothing it defeats most detergents and deodorants. For any animal that has once experienced the skunk's powers, the warning coloration of its fur is a sufficient deterrent at any future encounter.

Like the skunks of the temperate forests, the red fox emerges at nightfall to search for food. Relying on stealth and cunning, the fox stalks small mammals such as voles, mice, and cottontails among the shadowy underbrush, usually taking the unlucky victims back to its underground den to be eaten at leisure. Although small mammals are the mainstay of their diet, foxes are omnivorous and may also prey on large insects, birds and birds' eggs, and even fruit. In denser forests, the more retiring American gray fox makes its home. Its diet is the same as that of the red fox, but it has the unique advantage among foxes of being able to climb trees in search of food and in order to escape pursuit.

Scuttling rapidly across the forest floor, keeping to the thickest ground cover and the deepest

The white-tailed deer prefers woodland or the forest edge, where it can graze the grass but soon escape among the trees for safety when threatened by any kind of danger.

The distinctive black and white markings of the striped skunk warn would-be predators that the animal is best left alone.

shadows, the little deer mice emerge at night to feed on seeds, berries and other fruits, and occasionally small invertebrates. By day they rest in snug underground nests lined with shredded leaves and bark for, being short-legged and squat, they are able to burrow with ease. Although they are preyed upon not only by foxes but also by countless hawks, owls, weasels and other predators, deer mice are the most numerous of the small herbivorous mammals of the American temperate forests. Their success in the face of such adversity is due to their ability to breed rapidly.

Deer mice breed at all times of the year. The majority of them live for only a few months, being killed by carnivores while they are still young and inexperienced. To the species as a whole this is not important for, provided some deer mice live long enough to realize their great breeding potential, the species survives.

Another prolific species that occurs in forests from New England to Mexico is the American, or Virginian, opossum, which is a marsupial, distantly allied to the kangaroos. Like the deer mice, the opossum owes its success partly to its capacity for rapid breeding. It is also very adaptable, feeding omnivorously both on the ground and in trees, where it climbs skillfully, using its tail as well as its limbs to keep a firm hold.

Where the temperate forests border on rivers and lakes, the raccoon makes its home in a hollow tree or a crevice in rock. It is well adapted to this waterside habitat, dabbling with its long, slender fingers beneath the surface of the water for its favorite food—fish frogs, and crayfish. Using its dexterous forepaws the raccoon also climbs with great agility despite its rather stocky shape, searching in the trees for eggs, fruits, nuts, and seeds to supplement its diet. Being nocturnal the raccoon is easily overlooked although it is still common in many forests.

Forests offer a wider variety of ecological niches than does open country. One British observer of birds has, for example, estimated that on average there are up to five breeding pairs to the acre in woodland, whereas there are only two to the acre in permanent grassland. No doubt the numbers are a little lower in winter, for many birds that breed in the temperate forests migrate southward for the winter and their departure is only partially offset by the arrival of migrants from farther north.

Migration is a common feature in the lives of

Like all badgers, the American badger is a highly skilled burrower and excavates a large, very deep den known as a "set." Venturing from this den at night the badger usually hunts in grassland, digging small rodents from their burrows.

Deer mice are the most numerous of the small herbivorous mammals of the American temperate forests. Although preyed upon by countless birds and animals, they continue to thrive on account of their rapid breeding from spring until the start of winter.

Young Virginian opossums are carried on their mother's back after they have left her pouch. Here, an adult female and four young, out on a nocturnal expedition to forage for fruit, insects, and other food, rest in the forked branch of a dogwood tree.

Dabbling in water the raccoon catches its favorite food—crayfish. It also preys on other aquatic animals, such as frogs and fish.

Young opossums are suckled at their mother's pouch. They remain in the pouch until they are about 14 weeks old.

A formation of bean geese flying south from their Arctic breeding grounds to overwinter in the southern states of North America.

the birds of the temperate forests. In North America, for example, the migrants' flyways—down the east coast and along the Mississippi River, to name just two of the five main routes—are clearly discernable in spring and fall. They are the setting for many remarkable feats of endurance and instinctive navigation. The performance of the ruby-throated hummingbird, which covers well over 2,000 miles, from eastern North America across the Gulf of Mexico to northern South America, flapping its wings over 50 times every second of the way, is perhaps the most remarkable of all. Physiologists have proved that these tiny creatures cannot possibly store enough energy in their diminutive bodies for so long a trip, but despite this they continue to do the impossible, twice yearly.

Among the birds resident in the temperate forests throughout the year, some change their diet with the seasons in order to have enough food to survive the winter. In the fall the European song thrush, for example, feeds on small fruits, but during the short, hard days of winter it searches for hibernating snails and hammers them open, using a stone as an anvil.

Although the power of flight gives birds a vastly greater range than that of typical ground-dwelling animals, there are some ecological niches that they cannot fill. For instance, they rely on their eyes as their main sense organs, so they are not capable of catching great numbers of night-flying insects, such as moths. By contrast, insectivorous bats have very inefficient eyes. Instead, they use an echo-location technique with very high-pitched sounds that give a usable echo at a range of several feet. In this way

they can avoid obstacles and locate their prey in total darkness. In autumn, when most of the insectivorous birds migrate, some bats also migrate, but others hibernate, spending the winter almost lifeless until the warmer days and the insects return.

In the winter when such insect-eaters as the warblers and the nightingale of Europe have flown southward, and others such as some bats have gone into hibernation, some remain active in the temperate forests. The European robin continues to hold a territory throughout the year. Because of this the robin's song can be heard even in the short days of December and January when almost all other birds are silent. The song is a call signifying possession and defiance, and is essential if the territory is to be held. Some of the forests' seed-eating finches, on the other hand, abandon their territorial isolation with the coming of winter and gather in flocks, ranging more widely and feeding together.

As the seasons change and some of the hunted animals migrate or hibernate, the numbers of hunters must adjust accordingly, so some of them migrate, too. Some sparrowhawks, for example, the most typical predatory birds of the thicker temperate forests of Europe, migrate southward in the winter with their prey, thus maintaining the vital balance between hunter and hunted.

Migration over any great distance is less usual among the amphibians and reptiles of the forest, because they are incapable of sustained muscular activity. Nevertheless, some cold-blooded vertebrates do migrate over relatively long distances—the timber rattlesnake, for example, may cover distances of up to 25 miles. The European com-

An adult green woodpecker brings back food for its young, which nestle safely in a hollowed-out old tree. Woodpeckers are tree-climbing birds and can cling securely to tree trunks with their needle-sharp claws, using their stiff tail feathers as a prop.

Right: a common summer resident in temperate woodlands, thickets, and hedgerows, the chaffinch breeds throughout the European countryside. When not nesting, chaffinches are gregarious birds and they move about in large flocks in winter.

Bats can fly as well as birds, and some, such as this mouse-eared bat, fly even better than some birds. Venturing out at night in search of insects, it relies on an echo-location technique to find its way in the darkness. Like most bats, it emits ultrasonic squeaks through the open mouth and can assess the position of objects in its path by noting the delay between the squeak and the return of the echo. The pointed fleshy lobes or earlets, which are situated just in front of the ears, are important in this echo-location.

Left: the common jay of the temperate Eurasian woodlands is a distinctively marked and widespread bird. It feeds on a varied diet that includes acorns, beechnuts, berries, insects, and eggs, and is a permanent resident throughout its range.

mon frog also migrates, but only locally. By the time its food supply of insects and other invertebrates becomes scarce in the fall, the frog has developed a food-store of fat inside its own body. Leaving the damp parts of the forest where it lives during the summer, it makes its way to a pond, where it hibernates in the soft mud under the water. Toads, having skins that are more resistant to water-loss, hibernate in dry banks, often using the old burrows of small mammals. Small lizards and snakes may also hibernate close by.

In summer the reptiles of the cooler temperate forests are usually found in clearings, where sunlight can reach the ground. By moving in and out of the sun they can keep warm, and their body temperature can in this way remain quite steady under summer conditions. No doubt it is partly for this reason that in Europe the common lizard and the adder give birth to live young, the female retaining the eggs inside her body until they hatch. The grass snake has an alternative method of providing sufficient warmth for its eggs in cool climates. The female lays her eggs among decaying vegetation and the heat produced by the decay keeps the eggs warm and enables them to

A grass snake sidles over her batch of eggs. In the cool climate of the European deciduous forests, incubating eggs presents something of a problem for cold-blooded reptiles. The female grass snake lays her eggs in rotting vegetation, where the heat produced by decay keeps them warm until they hatch.

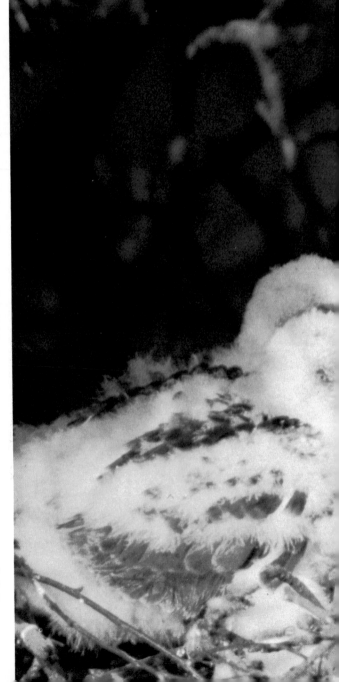

A sparrowhawk tears up food for its hungry chicks. The sparrowhawk's broad wings and long tail enable it to maneuver superbly among the trees in pursuit of the small mammals and birds on which it feeds. In winter some European sparrowhawks migrate southward, following the migration of their prey.

hatch. Suitable sites for egg-laying are not common, and many females may use the same site.

Living on and below the ground of the temperate forests of Europe is the wood mouse, which has a life style resembling that of the deer mouse of North America. It is active throughout the year, feeding on buds and insects in spring and early summer, and seeds and fruits, including nuts, later in the year. Wood mice are the most numerous of the small rodents of the forest floor within their range, but despite this they are not often seen, for they are nocturnal, spending the day in underground burrows. The bank voles of the forest floor are seen more frequently, for although they are most active at dusk, they may emerge from their nests during the day. Both mice and voles form an important part of the diet of predators such as owls and weasels.

Between the ground and the interlaced branches of the trees, in places where sufficient light filters through, shrubs form a distinct layer of vegetation. In Europe this shrub layer, which is common at the forest edges and around clearings, is the domain of dormice, small squirrel-like mice that climb adeptly. Dormice are omnivorous, eating seeds, bark, insects, and some-

times birds' eggs. Despite their varied diet, such food is not constantly available in the shrub layer. In addition, shrubs are more open to the icy blasts of winter than is the ground, which is protected by leaf-litter, and this is probably why dormice, unlike their ground-dwelling relatives, are true hibernators.

The common dormouse is the best climber for it is small and can rotate its hands and feet more than any other dormouse so it can grasp slender branches at a variety of angles. The edible dormouse, so called because the Romans used to eat it preserved in honey, is the largest of the dormice and the most squirrel-like in appearance and has a rather similar range to the common dormouse. It was introduced into a small area on the edge of the Chiltern Hills in Britain about 70 years ago and since then has neither died out, nor extended its range greatly. This restricted distribution shows that the edible dormouse has found a suitable localized ecological niche, but that the natural balance of forces is such that, in Britain as a whole, that niche is either precariously small or has other occupants.

Other animals introduced by man have produced more decisive patterns of settlement. The American gray squirrel introduced to Britain has filled the niche in deciduous forests once occupied by the red squirrel, and has become common within a century. The rabbit probably originally

The edible dormouse of Europe resembles a small squirrel with its thick, bushy tail and arboreal way of life. In Britain it fills a single, small ecological niche on the edge of the Chiltern Hills in southeast England. This restricted distribution would seem to indicate either that the niche is precariously small in Britain or else that it has other occupants.

The dormouse falls asleep at the beginning of winter and its body temperature drops rapidly to a level just above that of the environment. If the weather becomes very cold the dormouse increases its heat production so that its body temperature is maintained above freezing point. During warmer spells the animal tends to wake up and its body temperature rapidly returns to normal.

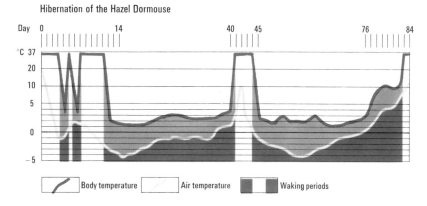

Hibernation of the Hazel Dormouse

Day 0   14   40 45   76   84

°C 37
20
10
5

0

−5

Body temperature   Air temperature   Waking periods

came from Spain, but was widely introduced elsewhere perhaps a thousand years ago as a source of human food. It rapidly became established and by feeding on seedling trees and gnawing on bark, has had a devastating effect on the environment, contributing greatly to the deforestation of many areas.

Just as some animals have been introduced and spread to new areas by man so others have suffered and become rare as a result of his actions. For example, the beech marten, an arboreal, squirrel-hunting weasel of the temperate forests of Europe and Asia has become rare in many areas, as has the ground-dwelling polecat. Both of these species are rather bold and inquiring by nature, and these qualities have been their undoing. The stoat and the smaller weasel are more cautious and have fared better. So, too, has the badger of Europe and Asia, which hides in a large underground burrow during daylight. Because the badger rarely kills chickens and game birds, it is generally tolerated by man.

In Europe and much of Asia potentially dangerous predators such as wolves and bears are now rare, and survive only in remote areas. There are still a few wolves in Scandinavia, central France, Spain, Italy, and the Balkans, and brown bears have an even more limited distribution in Europe. They, and other predators like the wild cat are regarded as too much of a threat to be tolerated, and their numbers are controlled by man.

The smaller and often less conspicuous insectivorous mammals have fared better than the larger carnivores. Shrews are still common, and the hedgehog has become adapted to coexist with human civilization and is often seen shuffling

Forming an important part of the diet of predators such as owls and weasels, bank voles are very common throughout Europe. Like all voles they are prone to cyclic fluctuations in numbers, and these cause fluctuations in the numbers of the predator species.

along in suburban gardens at dusk. Unlike its relatives the shrews and moles, the hedgehog is unable to pursue its prey beneath the ground. Invertebrates are not sufficiently common on the surface of the ground to provide it with an adequate winter food supply, and this is probably the reason for the hedgehog's hibernation. The red fox, too, has become adapted to man, and is thriving in many suburbs, sometimes even killing domestic animals in addition to its natural prey—rabbits, rats, voles, and birds.

Having reduced the number of larger carnivores that inhabit temperate forests, man has replaced these animals as the chief factor in controlling numbers. For example the red deer, where it inhabits parkland and private estates in Europe and Asia, is prevented from overcrowding its habitat by controlled hunting.

Red deer live in herds in the more open temperate forests. For much of the year the stags live apart from the hinds and young, but in the autumn they fight in order to gain control of harems during the relatively short mating season or rut. In parts of Britain red deer now live on bare moorlands, but the natural habitat of this species is forest.

Living in denser woods and thickets are the more solitary and much smaller roe deer. During the rut, males of this species mark out territories by means of the scent glands on their foreheads, thrashing at overhanging leaves with their short antlers. When a female appears, a highly ritualized courtship takes place in the form of a chase, which often leaves a well-worn circular track around a prominent tree.

The fallow deer was probably once found only in forests close to the shore of the Mediterranean Sea, but is now far more widespread. It is thought to have been introduced to many parts of Europe by the Romans. Like all deer, fallow deer are able to chew the cud. This means that they chew their food, usually leaves, twice. After being chewed and swallowed, the food is later regurgitated in a partly digested form and chewed again before

The European hedgehog, which feeds mainly on insects, snails, slugs, and other invertebrates, is a valuable pest-killer. As such it has survived well alongside the ever-growing human population and often makes its home in hedgerows and gardens in suburban areas.

The distinctively marked Eurasian badger is one of the few larger forest-dwelling carnivores to have survived the most ruthless predator of all—man.

Below: a red fox emerges from under-brush to drink at a stream. Hunting at night the fox relies on stealth and cunning to overcome prey. It sometimes becomes adapted to suburban life and may kill domestic animals in addition to its natural prey of mice and rabbits.

Squat in outline and protected by a thick skin, the once plentiful wild boar is well adapted for running through thick underbrush in the few wooded areas of Europe and Asia where it still exists. Wild boar are generally sociable and live in herds consisting of females and young under the protection of an old sow. At birth the young are striped with white, but they lose these markings after six months and become a uniform brown color.

Once widespread in forested regions, the European bison or wisent now survives only in zoos and in special forest reserves in Poland. It has a smaller head and longer legs and is altogether more graceful than its close relative, the prairie-roaming bison or "buffalo" of North America, which is also now rare.

In denser woods and thickets the small roe deer (left) is still common. Unlike the antlers of the red deer, which are large and extensively branched when fully grown, those of the roe buck are short and almost vertical.

Below: red deer are fairly common in more open wooded country throughout most of Europe despite being extensively hunted.

being swallowed for the second time. The advantages of this seemingly complex process are twofold; on the one hand, the animals can make more efficient use of plant food and on the other, they are able to gather food speedily and then retreat to a situation of greater safety to chew and digest their meal.

While deer still flourish in some European forests as a result of human interest, the largest forest herbivore, the European bison, or wisent, was nearly made extinct by man. The European bison is closely related to the American bison, but is taller at the shoulder and has a smaller head. Until the 1500s it was widely distributed in Europe, but the last unprotected member of the species was shot in the Caucasus Mountains of southern Russia in 1921. The only survivors now live in zoos, and in special forest reserves in Poland.

Another hoofed animal of the temperate forests that was once common, but which is now extinct over much of its former range, especially in western Europe, is the wild boar. An animal that

The bamboo forests of southwest China are the home of the solitary giant panda. Tender young bamboo shoots provide the panda with a ready source of food, and its front feet have become adapted to hold this diet by developing a thumblike structure.

runs through thick undergrowth must be squat and the wild boar is just the right shape to crash through the shrubs of the forest floor, passing underneath the overhanging branches.

Wild boar often feed on fruits and seeds that they find lying on the ground, but much of the food that they like best lies beneath ground level. Using their keen sense of smell, wild boar can find nourishing underground roots and fungi and also insect grubs. Rooting with their noses, they can easily dig up this buried food. Although they belong to a group of animals that are mainly herbivorous, wild boar are themselves omni-

vorous. By contrast, some omnivores, such as bears, are carnivorous in origin. From the extremes as either meat or plant eaters, some animals have developed toward an intermediate position as omnivores, their back teeth becoming rather like those of man with bumpy surfaces able to grind both plant material and meat.

The dense mountain forests of southwest China and the similar forests of southern Japan differ markedly from the cooler temperate forests we have already looked at. Here, broadleaved evergreens on the foothills merge with bamboo forest farther up the slopes and with swampy forests on

70

Japanese macaques play among trees in the deciduous forests of southern Japan. These heavily built, very long-haired monkeys are remarkable because they can survive cold winters that would almost certainly kill any other monkey.

A tree goanna clings expertly to a tree trunk. These lizards prey on a wide range of animals, from insects to small mammals, and they are fond of bird's eggs.

A yellow rosella, a common bird of Australia's eucalyptus forests.

A male superb lyrebird shows off his tail feathers in a mating display.

the lowlands of eastern China. The swampy forests are inhabited by the raccoon-dog, a small, short-legged animal that looks rather like the North American raccoon. Hunting at night, the raccoon-dog usually roams for considerable distances in search of frogs, shellfish, fish, and insects to feed on.

The bamboo forests are the home of the giant panda, a solitary animal whose main source of food is the young bamboo shoots among which it lives. Bones in the panda's wrist have become adapted to support a structure that looks and works rather like a thumb, and this is used to hold the food. Being heavily-built, the panda does not climb well, and for this reason it has only a short tail.

One of the most remarkable animals of the

A kookaburra in flight. One of the largest kingfishers, the kookaburra is renowned for its laughlike call.

forests of eastern Asia is the Japanese macaque, a heavily built, short-tailed monkey Almost all monkeys live in or near the tropics, but the Japanese macaques live in troops in deciduous forests, and in mid-winter they survive conditions that would almost certainly kill any other monkey. For two months or more each year, the forests in which they live are bare and often snow-covered. At this time Japanese macaques feed on a meager diet of bark, tiny buds, and any roots that they can dig from the frozen ground. Their shaggy coats keep out the cold.

The warm temperate forests that fringe the south-eastern coastline of Australia are the home of many unusual animals. Here, isolated on the island continent from more modern competitors, pouched mammals, or *marsupials*, still play a wide range of roles. Among the eucalyptus, or gum trees, rat-sized bandicoots and smaller marsupial "mice" feed omnivorously and scurry on the ground. Small members of the kangaroo family fill niches elsewhere occupied by small deer, and feed on the plants of the thicker forests, while larger kangaroos replace larger deer in the more open forests. Members of the phalanger family ("possums" to an Australian, but not to be confused with American opossums, which belong to a different family of marsupials) climb actively in the trees. Some phalangers have even paralleled the evolution of the flying squirrels, and can glide from one tree to another. Wombats burrow, and live rather badgerlike lives, and their closest relatives, the koalas, use their needle-sharp claws to climb gum trees in search

Koalas are among the most expert climbers of Australia's pouched mammals. The hands and feet make a firm clamp with three of the clawed toes opposing the other two. The name "koala" means "no drink," and is very appropriate for these animals because they gain sufficient moisture from the two to three pounds of eucalyptus leaves they eat every day.

Left: at dusk and daybreak the echidna, an egg-laying mammal, comes out of hiding to dig for food such as ants and other insects.

Above: the carnivorous mainly nocturnal, dasyures of Australia's temperate forests fill similar niches to those filled by the dogs, weasels, and civets of other continents.

Above: at home in both forest and open country, the Tasmanian devil is one of the largest carnivorous marsupials. Once found throughout Australia, it now lives only in Tasmania.

of the leaves on which they feed exclusively.

Among the carnivorous marsupials of the forest and underbrush are the small, weasel-like dasyures, which prey on birds and smaller mammals, and the larger, superficially doglike Tasmanian devil, which hunts lizards, wallabies, and large birds. The Tasmanian devil is also among the hunters of the more open bush country where the largest marsupial carnivore, the thylacine or Tasmanian tiger, made its home. The thylacine was once the fiercest of Australia's predators, able to hunt and kill the largest kangaroos, but it suffered as a result of competition from the dingo, introduced by man, and became extinct on the mainland. It is possible that it still survives in Tasmania.

Owing to the continent's isolation, Australia's temperate forests are the home of one species of another ancient and even more extraordinary group, the *monotremes* or egg-laying mammals. The echidna or spiny-anteater has hair and feeds its young on milk but, like the reptiles, it lays eggs. Despite its ancient features, the echidna presents a modern and efficient exterior to the world. As it scurries through the forests at night it is protected from enemies by its stiff, spiny hairs. It rips open ants' nests with its powerful claws, and feeds on the exposed insects using its beaklike face and its long, thin, sticky tongue.

Even though seas are less of a barrier to birds than animals, the birds of Australia's temperate forests are, like the mammals, unusual. Cockatoos—large, crested birds belonging to the parrot family—are found only in the forests of Australasia and the neighboring islands off the mainland of Asia. The smaller rosellas and the nectar-loving lories may also be seen in the eucalyptus trees along with noisy kookaburras. Well-known for their call, which sounds like a human laugh, the kookaburras belong to the kingfisher family, but instead of using their long bills to hold slippery fish they feed mainly on large insects, small lizards, and snakes.

Also resident in these forests are the lyrebirds. The male superb lyrebird has one of the most magnificent displays of tail feathers seen in birds, being able to sweep the long feathers forward in a beautiful arch over his body. Each male establishes a large territory up to half a mile across in thick forests, and within this defended area he constructs several low mounds of soil on which he displays daily. His display song includes quotes from those of many other species of birds, for lyrebirds are perhaps the most accomplished of all mimics. As is often the case in species in which the appearance of males is particularly striking, lyrebirds are polygamous. The splendid males take no part in nest-building, incubating, or caring for the young, leaving these activities to their mates.

In New Zealand, isolation of the islands resulted in the absence of ground-living mammals, and consequently, living space on the ground, which elsewhere is occupied by mammals, was available to birds. Beyond this, because there were no predatory mammals on the ground, large birds had no need to fly in order to escape from danger, so when large birds—sometimes too large and heavy to fly—evolved in New Zealand there were safe ecological niches for them to fill. Until man arrived on the islands and started to destroy the forests, the country provided a paradise for flightless birds.

Living in the thickest of the forests are the small shy kiwis, which have only tiny wings and no visible tail. They are active at night, when the earthworms that they eat are most likely to be near the surface. Unlike any other birds, kiwis have nostrils near the tip of the beak rather than at its base. They hunt earthworms by scent, snuffling along like feathered bloodhounds, with their nostrils close to the ground.

Another nocturnal bird of New Zealand, the kakapo, is now rare, but a few still survive in the mountain forests of South Island. Being heavily built, and with only weak wing muscles, the kakapo is unable to undertake powered flight but it sometimes climbs trees to obtain fruit and nectar, and then glides back to the ground.

Spending the summer in high, rushy mountain valleys is the rare, flightless takahe, which, like the kakapo, is unique to New Zealand. The takahe migrates to the forests nearby for the southern winter. The related weka is another flightless bird, but provides an unusual example of survival in the face of man, and the predators that he has brought with him. Being carnivorous, the weka finds the recently introduced rats and mice a welcome addition to its diet. Also known as the wood rail, the weka is a brownish bird about the size of a chicken.

Above: the shy, flightless kiwi of New Zealand uses its long beak to probe the ground for earthworms during the night. As more and more of its forest habitat is cleared by man, the kiwi is being forced to adapt to life in scrub and forest fringes.

Left: the kakapo is a rare parrot of New Zealand's mountain forests. It is particularly vulnerable to introduced predators because it has lost the power of sustained flight. It climbs trees to reach fruit and nectar but can only glide back to the ground.

The takahe of New Zealand, a flightless bird once thought to be extinct but recently rediscovered living in South Island, summers in high, mountain valleys and winters in beech forests lower down. It is about the size of a large domestic hen.

A Simple Food Web in a Temperate Forest

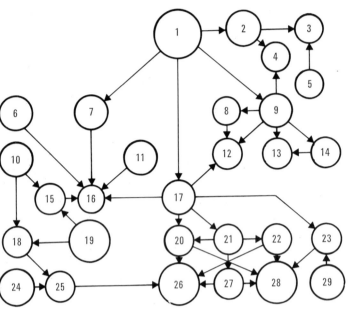

1 Broad-winged hawk
2 Blackburnian warbler
3 Winter moth caterpillar
4 Aphid
5 Common shield bug
6 Gray squirrel
7 Rufous-sided towhee
8 Ophion (parasite)
9 American redstart
10 Barred owl
11 White-breasted nuthatch
12 Mottled umber moth caterpillar
13 Gall wasp
14 Braconid wasp (parasite)
15 Woodmouse

16 Acorns, Beechnuts
17 Tufted titmouse
18 Common shrew
19 Gray fox
20 Springtail
21 Wolf spider
22 Millipede
23 Beetle larva
24 Woodthrush
25 Earthworm
26 Litter (Fungi and bacteria)
27 Mite
28 Dead wood
   (Fungi and bacteria)
29 Hairy woodpecker

This deciduous forest food web is typical of the natural hardwood forests of eastern North America, but also similar to that of the extensive forests of Europe. The diversity of trees is greater than in coniferous forests, some of the commonest dominants being oak, beech, and maple, with poplars, sycamore, and willow occurring largely in the wetter areas. The most obvious feature of these forests is the absence of leaves during winter and early spring. This allows sufficient light for a rich field layer of plants, many of which have a growing season confined to the early spring before the trees come into full leaf. The flush of tender leaves in spring provides a rich source of food for innumerable kinds of insects, largely caterpillars and plant bugs. These in turn are the main source of food for many other animals, especially insectivorous birds such as the warblers, which migrate from the South to breed amid the plentiful food supply.

The increased diversity of animals compared with a coniferous forest is a direct result of the greater richness and structural diversity of the plant community. The number of links in the total food web is also much increased, which in turn tends to limit the fluctuations in the size of individual populations.

Tropical Forests

In the tropics the sunlight is stronger and more constant than in any other part of the world. Where the rainfall is only moderate, the lack of water sets a limit on the vegetation, and at best only thorny trees, which may shed their leaves during the driest part of the year, can grow. If the rainfall is abundant but seasonal, the result is a more luxuriant monsoon forest. Heavy rain at all times of the year makes possible the most remarkable of all forests, the tropical rain forests, where in both numbers and productivity the earth's plant life reaches a crescendo.

In Central and South America, in equatorial Africa, in Southeast Asia and near the northern coasts of Australia where there are from 40 to over 100 inches of rain each year, giant trees tower 160 feet into the air, their bushy crowns standing out like islands over a green sea of the foliage that decks the slightly shorter trees. Enough light filters through this dense green canopy to permit the growth of more scattered shorter trees and shrubs. On the leaves, branches, and trunks of the larger plants grow *epiphytes*—smaller plants that manage to take root in the crevices of trees and climb to where the light is brighter than it is on the ground below. Climbing plants rooted in the ground, use trees for support, their stems—often thicker than a man's arm—reaching through the green canopy toward the light. Where a giant tree has fallen, herbs flourish briefly on the ground before rapidly growing branches and leaves once again blot out the sky. The variety of plant life is astonishing. In the average temperate forest there are about 12 species of trees; in a tropical rain forest there may be over 400.

Most trees of the tropical rain forest are evergreen with dark green leathery leaves that fall when they become old, and are replaced at all times of the year. Some trees shed their leaves all at once at irregular intervals. However, there are regular seasonal changes, particularly when it comes to flowering and fruiting. This means that

Rivers like this, the Suiá Missú, wind through many of South America's tropical forests. Although conditions are ideal for growth, the maze of waterways can limit species to small areas.

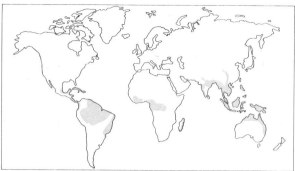

The World's Tropical Forests

The map at left shows the areas of the world's tropical forests. At the foot of this page are drawings of trees typical of the forests of equatorial Africa, to the same scale as similar drawings on pages 18 and 44; note how much taller they are than forest trees of cooler climes. Below are details of their foliage. The deer on the facing page are axis spotted deer, or chital, commonly found in the forests and forest margins of India and Sri Lanka, where they live in large herds. Their spotted flanks give excellent camouflage in the forest shade, where sunlight filters through the canopy of leaves to produce a very similar effect.

although the tropical forest provides large amounts of food for nectar- and fruit-eating animals, the supply fluctuates, and these animals may have to range quite widely during the year in order to survive.

The towering vegetation of the forest shields the tangled spaces below from the heat of the sun and from the wind. Beneath the canopy the temperature remains steady, and the humidity is high and constant. Of all land environments this is the most even and unchanging; it therefore poses the fewest problems to animals. It is not surprising that in terms of numbers of species and of individuals the tropical rain forests support such an incredibly rich animal life.

Only in the soil is animal life relatively scarce. This is because the constant warmth and mois-

ture ensure that complete decay of fallen leaves is rapid, with the result that the soil is not very rich in nourishing humus as a source of animal food. So great is the biological productivity of the forest, and so rapid is the turnover of used plant materials that little remains to support life in the soil. In compensation the leaf-litter on the surface of the ground, and the sheltered crevices among the roots of the clinging epiphytic plants teem with life.

In the moist air thrive animals belonging to groups that elsewhere live only in water. Brilliantly colored flatworms up to 20 inches long glide over moist leaves in search of their prey. Leeches cling inconspicuously, ready to become attached to, and then suck blood from, the first larger animal that passes. Moisture-loving land

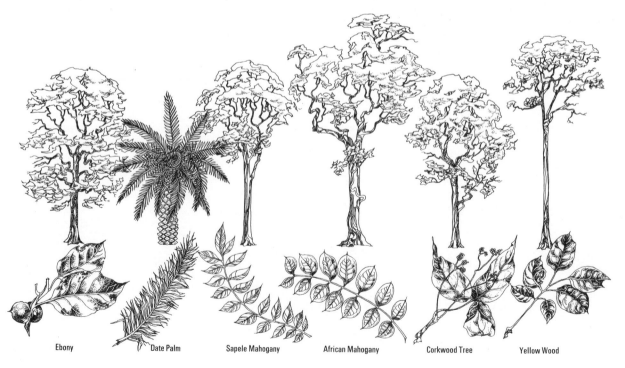

Ebony   Date Palm   Sapele Mahogany   African Mahogany   Corkwood Tree   Yellow Wood

Confronted by a forest millipede, the natural reaction of this velvet worm is to discharge a milky fluid that congeals to entangle the prey in a mesh of sticky threads, leaving it helpless.

Looking rather like a twig as it clings among leaves, this leech is ready to attach itself to any passing animal for a blood meal.

Above right: this jumping spider, here shown pouncing on an unsuspecting fly, captures its prey by stalking it slowly.

animals are also common and, under such ideal conditions, often reach almost incredible sizes. Giant snails and giant millipedes feed on the leaves, and giant centipedes feed carnivorously. There are giant spiders and scorpions, too, as well as a host of smaller ones.

Also flourishing in the moist environment are velvet worms that may be up to six inches long. These avoid strong light, spending much of their time, often in groups, beneath rotting fallen trees. In self-defense they can discharge a sticky substance from openings on either side of the mouth at any intruder. They feed on small spiders and insects, and are armed with sharp, pointed jaws.

The tropical rain forest teems with insects. Tiny drably colored wingless insects abound in the leaf-litter on the ground. By feeding on the

decaying vegetation they rapidly return important minerals to circulation so that they can be used again by other living things, and to a large extent fulfill the role carried out by earthworms and other soil dwellers in temperate forests. Also important in breaking down dead plants are the termites. No dead tree is long without its colony of termites, excavating galleries to form an extensive nest, and feeding on the dead wood. Termites are able to digest wood because they have colonies of tiny single-celled animals—protozoans—living within their intestines. Wood swallowed by the termites is broken down by the

Right: in a tropical forest in southern Africa an orb-web spider waits for the next victim to become ensnared in its intricate web. Many spiders instinctively build webs like this one.

protozoa into a form in which it can be absorbed and used by the termites.

Ants also live in huge colonies. Though they occur in other regions, they are nowhere as numerous as in the tropics. Within a colony there are types of individuals with their own separate functions—breeding, gathering food, or in some cases fighting. In places, streams of ants swarm over the forest floor, attacking and devouring almost any animal they encounter. Other species live more peaceably on fungi that they cultivate for themselves. Some colonies of ants roam continuously, carrying their larvae and pupae with them; others construct nests, sometimes by joining large leaves together with silk, which is produced by their larvae.

Plant-eating beetles are numerous, too. The world's largest insect, the goliath beetle, which is as large as a pack of cigarettes, is an inhabitant of the African forests. Large and iridescent dragonflies flit to and fro in clearings, feeding on the small midges that dance in swarms. Other insects live higher in the forest canopy, and are less often seen from the ground. High up, out in the sunlight, are the flowers of some of the forest trees. Here, gathering and feeding on the nectar, are bees and brilliantly colored butterflies and moths. Once again the rain forest proves to be the home of giants, for the largest butterflies have a wingspan of 10 inches, and the biggest moths are larger still.

The warm moist environment and abundant invertebrate life provide the ideal conditions for frogs and toads, and some of these are giants, too. The world's largest frog is the goliath frog, an inhabitant of some African forests and more than 13 inches long in the body, while the world's largest toad, which may grow over nine inches long, is found in the forests of Colombia and Ecuador in South America. Other species nearly as large live on the moist floor of forests in Brazil and Malaysia.

Tree frogs are also common in the tropical forests but they are smaller because they must be able to support their own weight while clinging to leaves or branches by means of the suction pads on their toes. Many tree frogs can change color so as to be camouflaged against any natural background, but some of the South American frogs have poisonous glands in their skins as a defense against enemies, and are brightly colored so as to warn off attackers.

Pools of still, fresh water are not common in

This delicate nest is the work of the ingenious tailor ants, which use their own silk-producing larvae as tools. While leaves still attached to the tree are held in place, the larvae are passed to and fro so that their silk binds the leaf edges together.

Largest and heaviest of all insects is the plant-eating goliath beetle of the African forests, seen here on a bunch of bananas. Some specimens of this beetle are over five inches long.

Brightly colored morpho butterflies swarm by a river in the Orinoco region of Venezuela. While the female butterflies remain in the depths of the tropical forests soaring among the treetops, the males often gather in hundreds beside rivers or mineral springs. It is thought that they may be attracted to the water by the presence of dissolved salts.

Carrying their larvae and pupae, safari or driver ants stream across the forest floor in huge columns. The fiercest of the insects of the tropical forests, the large soldier ants will kill almost any animal lying in their path. They often attack termite nests, killing the termites with poisonous bites and then tearing them to pieces so that the smaller worker ants can carry the prey away.

Above: although caecilians spend most of their lives underground, this one has emerged to eat an earthworm.

Above: the South American giant toad is among the largest of the world's toads, and may grow to as much as 10 inches in length. Because of its appetite for insects, it has been widely used in the control of agricultural insect pests.

tropical forests, and since amphibian tadpoles would be swept away if the eggs were laid in streams and rivers with a strong current, there is a shortage of conventional amphibian breeding sites. As a consequence of this problem the forest-dwelling frogs show a range of breeding adaptations without parallel elsewhere. Some tree frogs attach their eggs to leaves overhanging rivers, so that only the tadpoles that emerge from them and are able to swim have to contend with the river current. Others lay their eggs in the pockets of water that collect in rotting holes in trees, or even in the cup formed where large leaves join the stem of a plant. Some frogs lay eggs from which miniature, but perfectly formed frogs eventually emerge in the damp leaf-litter or the topsoil immediately below it. Another method of producing the same result, although a little more slowly, is shown by the South American frogs that deposit the eggs together with a foamy mass of froth, in the soil. The froth keeps the tadpoles moist until they turn into tiny frogs. The fertilized eggs of the "marsupial" frog are carried inside a moist pouch on the mother's back where they develop into fully formed froglets before they emerge.

Burrowing in leaf-litter or moist soil and seldom seen are the caecilians—legless, wormlike amphibians with small, often useless eyes. Most of them are drably colored, which is understandable in animals that spend almost all of their time underground, but surprisingly, a few species are brightly colored, having a vivid blue skin, or bright yellow stripes down the sides of the body. Their feeding habits are not well known, but it seems probable that they prey on insects and other invertebrates. Some caecilians give birth to live young, and others lay their eggs in the soil. In some cases the tadpoles wriggle their way to water nearby when they hatch, but by then they have usually lost their gills, so even if they spend some time in water they usually have to surface to breathe.

The warm environment of the tropical forests is as suitable for reptiles as it is for amphibians, and both on the ground and in the branches lizards and snakes flourish. Those of the South

American forests are particularly unusual, because for much of its history South America has been a huge island, unconnected with any other continent. Because of this isolation the reptiles (and other animals) of South America are excelled in diversity and strangeness only by those of Australasia. Nowhere else, for instance, are the long-tailed lizards called iguanas so plentiful and diverse. In fact they are found only in the Americas and in a few isolated islands such as Madagascar and Fiji. Elsewhere in the tropics their niche is filled by another lizard family of rather similar appearance and habits, the agamids.

Being active during the day, iguanas have keen eyesight, and they usually display prominent and often colorful visual signals during courtship, or as a threat when defending a territory against intruders. For example, the common iguana, a large tree-climbing lizard of tropical American forests, has a prominent crest of leatherlike spines running along its back. Unlike most lizards, iguanas are mainly herbivorous. The common iguana feeds mainly on fruits, tender leaves, and flowers, although some small animals also form part of its diet.

Belonging to the same family are the often brightly colored anoles or tree-lizards. Because they are small in size and have flattened adhesive pads on the ends of their toes the anoles climb very well. They fill the same sort of ecological niche as the African chameleons, and like the chameleons change color for camouflage against a variety of backgrounds, and also during courtship and when under stress. The related basilisks also live in the tropical forests of America. They climb well, using sharp claws instead of climbing pads. On the ground they run at speed on their hind legs alone, with their long tail held well off the ground to balance the head and body. So rapidly do they move that small basilisks have been seen skittering across water, their weight being supported as water supports a stone skimmed across its surface.

South America's forests are the home of a truly giant constricting snake, the anaconda, which, like all constrictors, kills its prey by squeezing it to death. The anaconda is never found far from the great forest rivers. Among thick branches, its huge, spotted body can be surprisingly inconspicuous as it lies digesting its last huge meal or waiting in ambush for the next. The longest anaconda ever measured was over

Left: in the depths of the forest recognition by sight is difficult, and male tree frogs, like this one, often make distinctive calls to find a mate. They are well adapted to living in trees and can grip branches and leaves with suction pads on their fingers and toes.

37 feet long, although the average is under 20 feet. This great variation is due to the fact that there is no such thing as a fully grown snake, for snakes continue growing as long as they live, molting their scaly skin at intervals.

Poisonous snakes are never as large as the biggest constrictors, for they rely on poison instead of weight and strength to overcome their prey. They are deadly enough despite their smaller size. Among those living in the tropical American forests are the bushmaster and the fer-de-lance, both equipped with deadly venom.

Because forests provide what is in many ways the most suitable of all environments for birds, and because the Central and South American forests are vast in area and varied in character—from low-lying riverside forest, to cooler, more open forests on mountain slopes—there is an almost incredible wealth of bird life. For instance, the many members of the parrot family fill the forests with their raucous voices as they squawk, chatter, burble, and shriek at each other. Flying among the trees in search of food, they add a wealth of color to the canopy with their bright and sometimes even gaudy plumage.

Most parrots are seed-eaters, for uniquely among the birds they have both the lower and the upper parts of the bill hinged against the skull. This gives them terrific biting power, enough to open even the hardest seeds. Largest

Above: a leaflike appearance has not saved this bush-cricket, or katydid, from falling prey to an anolis lizard. The katydid relies on mimicry for protection, but the anole is able to change color to blend with its surroundings.

Left: one of the most magnificent of boas, the emerald tree boa, rests on a tree branch, its brilliant green color blending with the surrounding foliage. The boa's coiled position enables it to pounce quickly when a victim approaches.

of all the tropical American parrots are the brightly colored, long-tailed macaws, at up to three feet long. They can deal with the largest and toughest of seeds, and are capable of cracking a Brazil nut, not with explosive force, but with a slow deliberation that requires greater controlled strength. As the tropical American parrots range widely in size, the variety of seeds available as food is fully exploited.

There are many species of both Amazon parrots and conures differing to human eyes only in small details of size and coloring. No doubt these differences are important to the birds for, when there is a possibility that wild animals may meet members of another, very similar, species there

must be some mechanism to ensure that they can recognize their own kind. It would not be of value to a species if many of its members wasted a breeding season courting and attempting to breed with members of another species and producing either infertile eggs or, at best, young hybrids probably lacking the useful and special adaptations of either parent species. It is significant that most of the colors and markings that distinguish the various species of Amazon parrots and conures are on the face and head, or on the part of the wing that corresponds to the human wrist and that is prominently bent forward when the wing is folded at rest. In these positions the signals by which parrots can in-

The incredible variety of birds in South America has led to it being called the "bird continent." Among the most familiar species are members of the parrot family, which, like the majority of South America's birds, live in the tropical forests. The parrots range widely in size and color, among the largest being the gold and blue macaws shown here, which measure over 30 inches from bill to tail-tip. Jays of brilliantly colored species are also found. They contrast markedly with their less gaudy relatives, the crows and ravens of temperate lands. Confined to America and most prolific in the tropical forests are the hummingbirds, whose dazzling iridescent colors often surpass even those of the birds of paradise found in New Guinea. There are over 300 species, ranging in size from the $2\frac{1}{4}$-inch bee hummingbird to the $8\frac{1}{2}$-inch giant hummingbird. All feed on the nectar of flowers and on small insects, using their long tongues and hovering flight to gather their food on the wing.

Hawk-headed parrot

Plush-crested jay

Green jay

American oriole

stinctively recognize members of their own species are most obvious.

The tropical American parrots present a fascinating situation to ecologists, for it is generally accepted that if two species compete for the same ecological niche in the same area, then one is almost certain to be more successful, and to thrive at the expense of the other. The abundance of parrots would seem to indicate that each species is adapted for a distinct way of life in the tropical forest, but we do not as yet know the precise nature of the niches they occupy.

Were it not for the sheer abundance of their species, the picture might be a little clearer in the case of the hummingbirds. The members of this family, found only in North and South America, all visit flowers where they feed on nectar, drinking the sweet syrup through their long, tubular tongues. Since flowers vary in shape and size, dif-

ferent species of hummingbirds have become adapted to feed on nectar from particular species of flowers with beaks of varying lengths and varying degrees of curvature.

Many hummingbirds are tiny. The smallest, which is also the smallest living bird, is the bee hummingbird, which builds a nest no larger than a thimble. Even in warm forests it is difficult for so small an animal to retain body heat during the night, when the bird's eyes do not work well and it cannot see clearly enough to search for food to replace the heat energy that is lost from the body. Because of this the body temperature of the smaller hummingbirds falls at night, so as to reduce the amount of heat lost and lessen the risk of starvation.

Partially competing with the hummingbirds for food are the slender-billed American honeycreepers. Piercing the bases of flowers with their

Yellow-napped parrot

Gold and blue macaws

Ruby-throated hummingbird

Glittering-bellied emerald hummingbird

Doctor bird

beaks, the honeycreepers use their brushlike tongues to obtain nectar. In this way they are able to get nectar from some flowers that would defeat many hummingbirds, but in other cases the hummingbirds have the advantage, being able to hover and so to feed from flowers inaccessible to the non-hovering honeycreepers. Because of their feeding habits, the honeycreepers are important as pollinators of some kinds of trees.

Birds that eat the larger fruits in South American forests belong to a group not found elsewhere, for although fruit-eating birds with similar adaptations are found in the tropical forests of Africa and Asia, they belong to a totally different family. The South American birds that fill this ecological niche are the toucans, which use their huge beaks for picking fruits from the tips of slender branches that would not bear their weight. Showing con-

siderable juggling skill, the toucans throw picked fruit from the front of the beak to the back so that they can swallow it. Although large in size these beaks are not as heavy as they look, for if they were, no toucan could fly. The outside, often brightly colored so as to make a very showy signal, is made of light, horny material, while the inside contains light, very spongy bone.

Smaller fruits are eaten by the tanagers— smaller, more conventional looking, often very brightly colored birds, and by the quetzals. Excelling the tanagers in color, the quetzal is probably the most striking of all the tropical American birds, being iridescent green with bright red feathers on the lower part of its breast. The male has a long, trailing tail of green feathers as remarkable in its way as that of the lyrebird of Australia. The quetzal lives a solitary life in the densest and darkest parts of the forests of

93

Easily recognized by their small heads with an untidy crest of feathers, hoatzins are suggestively ancient-looking birds with features that recall their reptilian ancestors: a musky odor like that of the crocodile, and a call that sounds more like a reptile than a bird. Not surprisingly, they are sometimes known as reptile-birds.

Central America. Darting from its perch in the trees, it feeds mainly on insects, which it can easily catch on the wing, for its downy plumage enables it to fly almost noiselessly. It also feeds by hovering to pick berries and small fruits from the trees.

Even more remarkable in some ways is the hoatzin, a leaf- and fruit-eater that lives in small flocks in forests usually not far from the banks of rivers. For the first two or three weeks of their life the chicks of this species, which become active as soon as they hatch, have two large claws at the front of each wing, with which they clamber through the branches. The claws are absent from adult hoatzins, but during their brief presence they suggest the sort of life that the birds' remote reptilian ancestors must have lived. However, the hoatzin is no "missing link." It has some suggestively ancient-looking features, certainly, but in other ways it is a modern bird.

Scratching the surface of the ground for grubs, insects, and seeds, the tinamous of the tropical forests of America both look and feed rather like chickens. Being large birds they fly only feebly if at all, and their whistled cries may be heard haunting the thick forest undergrowth.

Instead of the tiny hoofed antelopes and chevrotains that dash through the undergrowth of African and Asian forests, South America has long-legged rodents lurking in the thick forest cover. Before South America's long isolation some rodents had invaded the continent, and so became cut off there. With their large, grinding cheek teeth, they were for millions of years the most efficient herbivores to live there, for the hoofed mammals were kept out by the sea. Because of this, in South America the rodents came to fill ecological niches that elsewhere are occupied by hoofed mammals.

The acouchis are the smallest of these running rodents, being less than eight inches tall at the shoulder—slightly smaller than any hoofed animal. The longer legged agoutis have the advantage of being able to run faster, but are too

The female quetzal is noticeably less striking than her male counterpart, which is often claimed as the most beautiful of all the tropical American birds. Quetzals are rarely seen because they are solitary birds and live in the densest forests.

A toco toucan rests on a branch in the lowland forests of South America. Its brightly colored bill, which despite its size is very light, is well adapted to picking fruit from branches that are too slender and precarious to bear the toucan's weight.

tall to shelter in such thick cover. Like many other rodents, acouchis and agoutis often clasp their food between their forepaws, and sit erect to eat. The fact that they both store food when it is plentiful, burying it in the ground near some prominent landmark, is an indication that even in tropical forests the food supply can vary from one season to another. Close study has revealed that these rodents do not remember where their food stores are; when food is scarce they simply search in the most likely places, behavior that is not unknown in human beings who have mislaid something.

Agoutis and acouchis are usually active during the day. By night they sleep in shallow burrows, while their larger relative, the handsomely spotted paca, takes over the night shift. Pacas prefer to live in forests near water, and their favorite foods are fallen fruits such as avocados. In the same habitat, along the river banks, live the world's largest rodents—the size of sheep, but looking like huge guinea-pigs—the capybaras.

These graze on the grasses on land and the aquatic plants in the rivers, and, at the first sign of danger seek safety in the water.

Similarly, on account of their isolation, the tropical forests of the West Indies are the home of some unusual mammals. Unique to the islands are the ratlike hutias, which feed on grasses, shrubs, and trees. Some of them are mainly ground-living, but other species whose short, naked tails are prehensile, climb quite well. Sadly, these animals are now becoming rare because they have suffered in competition with the rats introduced accidentally by man, and also by falling prey to the mongooses introduced deliberately in a misguided attempt to control the rats. Also unique to the island's tropical forests are the solenodons, looking rather like giant shrews. Emerging at night from their shelter, they rely on their senses of smell and touch rather than on the sight of their tiny eyes to locate their food. Since they have no effective defense against predators, they are particularly vulnerable as they waddle clumsily along on the ground, in search of invertebrates, small reptiles, and fallen fruits. In the past this lack of defense was no disadvantage for the solenodons had no enemies. But by introducing rats, mongooses, dogs, and cats, man upset the balance, and solenodons are now, like hutias, dwindling toward rarity.

Despite the success of the rodents, the tropical forests of America are not entirely without hoofed mammals. Living close to forest rivers and lakes, for example, are the shy, solitary tapirs. These animals once flourished in many parts of the world but now survive only in tropical America and parts of Southeast Asia, where competition is in some ways less keen. In South America their only enemies are jaguars and caymans. Tapirs are large and heavily built with rather short legs, and can crash their way through thick vegetation. They feed by browsing

on water plants, and also on the lower branches of trees, using their long, movable upper lips and noses—like miniature versions of an elephant's trunk—to tuck the food into their mouths. South American tapirs are gray-brown in color, but their young have patterns of prominent white stripes and spots running the length of the body. Presumably these markings provide camouflage amid leafy shadows.

The American tropical forests have become the home of other more familiar hoofed mammals in addition to the tapirs. Among them, species of deer such as the white-tailed deer that also inhabits North America. In thicker forests lives the little brocket deer, while the larger swamp deer, in which the male of the species has long spreading antlers that would be an encumbrance in tangled undergrowth, keeps to the less dense vegetation near the forest fringes.

High in the spreading branches of the tropical

Being essentially aquatic in habit, capybaras seek safety in water at the first sign of approaching predators or other dangers.

forests of America lives a unique assembly of climbing mammals. There are opossums in variety, about as big as squirrels, feeding on insects, small vertebrates, and fruit. These opossums are marsupials and, apart from the North American opossum, are the only marsupials to survive outside Australasia. Surprisingly, some of the South American marsupials lack pouches. For example, the numerous young of the little murine opossum simply cling to the mother's teats and fur or, when they are a little older, cling to her back. The American opossums climb by means of their long, sharply-clawed toes, and have long, often naked tails that are useful in balancing and for wrapping around a branch in order to get a better grip.

Above: this young Brazilian tapir will lose its stripes and spots and gain the uniform dark brown color of an adult. The juvenile markings provide camouflage for the defenseless young among the forest shadows.

Left: along river banks in tropical America live the world's largest rodents, the capybaras. They eat grasses on land and aquatic plants in rivers, and their partially webbed feet make them well adapted to a riverside life.

The golden lion marmoset, which lives in the forests of the Amazon basin, is the most brightly colored of all mammals.

Clinging firmly to a branch by its tail and protected by its numerous barbed spines, the coendou, or Brazilian tree-porcupine is safe against almost any attack. It is the only porcupine to have a prehensile tail. Animals with prehensile tails are more common in South America than anywhere else. The kinkajou, a member of the raccoon family, is another example. It is an active nocturnal climber, feeding mainly on fruit, licking the juicy pulp with its long thin tongue. The olingo, similar in appearance and habits to the kinkajou, does not have a prehensile tail, however. At first sight this would seem to be a disadvantage, but it cannot be serious because the olingo survives. There must be some undiscovered compensating advantages.

Roaming through the forests, equally at home on the ground or up among the branches, troops of coatimundis forage for food. They put their long noses into every crevice, searching for tasty morsels—fruits, fungi, insects, or small or medium-sized vertebrates. Sometimes a large iguana, basking on a branch will drop to the ground for safety as the coatimundis approach, but this ruse is not always successful, for some members of the troop may remain on the ground in order to snap up such delicacies.

In the trees the coatimundis will sometimes meet a sloth—peaceably, for they will not come face to face, as coatimundis balance above the branches, while sloths hang below keeping a tight hold with their long hooked claws. Since these claws are effective weapons in self-defense, sloths have no need to beat a hasty retreat and, in fact, they climb only very slowly, sometimes even spending their whole lives in a single tree.

The tropical forests of South America are the home of a unique assembly of climbing mammals. Left: the three-toed sloth hangs upside down from branches using its huge claws as hooks. Center: the coendou, or tree-porcupine, has a prehensile tail well adapted for climbing trees. Right: the kinkajou's tail is a "fifth limb" that leaves its forepaws free to reach after succulent fruits.

Although sloths are quite large for tree-dwelling animals, they are often difficult to see, for their rough hair harbors tiny green algae, very similar to those that may grow on the trunks of trees. These algae make the sloth look green, and are a great help as camouflage among the leaves.

Smaller and much more active climbers, occasionally traveling in troops of a hundred or more, are the marmosets and tamarins. These are allied to the American monkeys, and indeed some people would include them in the monkey group. Marmosets are often strikingly marked, and shriek and chatter to each other very much as birds do. They feed on insects, shoots and fruits, and climb well, using their long tails for balance. Since marmosets are smaller than monkeys their fingers and toes are often too short to reach around the branches and secure a grip. Instead,

they use their nail-like claws to cling to the rough bark like squirrels.

Among the commonest true monkeys that inhabit the forests of Central and South America are the howlers. These monkeys live in family groups and defend a territory by shouting defiance at their neighbors. Howlers, like the lightly built spider monkeys and the burly woolly monkeys, have touch pads, rather like finger prints, on the underside of the prehensile tail near its tip, so as to ensure that the tail can grip firmly. The little capuchin monkeys also cling with their tails, but lack touch pads. The bald-headed uakaris, the squirrel monkeys, and the sakis use their tails simply as balancing organs, and cling to the branches only with their long fingers and toes.

As night falls in the American tropical forests,

Geoffroy's spider monkey lives in the densest forests of Central America. It swings, apelike, beneath the branches.

the world's only nocturnal monkey, the dourou-couli, begins its hunt for food. It fills broadly the same ecological niche in America as the larger bushbabies of Africa. Like bushbabies and owls, it has huge eyes, able to admit every possible glimmer of light, so it is able to search for its food—fruit, leaves, insects, and small birds—during the hours of darkness. The nocturnal climbing anteaters, such as the pygmy or two-toed anteater, and the slightly larger tamandua also become active as the daylight fades. Both these anteaters have prehensile tails, which they use to form a tripod with the powerfully clawed hind feet while they attack ants' nests with their forepaws. If threatened, these climbing anteaters rear up, holding the hooked claws of the fore-paws ready for use as weapons.

The evening is the most likely time for small packs of bush dogs to hunt in the undergrowth for the large rodents, such as the paca, on which they prey. Though a true member of the dog family, the bush dog has come to differ markedly from all other dogs during its long isolation. It is small, squat, and short-tailed, and looks rather like a heavily built weasel.

Of the many cats that prowl among the leafy shadows on the forest floor during the evening as well as during the day, the largest is the jaguar. Well camouflaged by its spotted coat, the jaguar often lurks near rivers preying on animals, such as deer and agoutis, that come to drink. Unlike some other cats, jaguars have no fear of water and swim well, so capybaras, which often seek safety in the water, may still not escape from them. Whereas in Asian forests the tiger is best adapted to hunt the largest prey, usually on the ground, while the slightly smaller and more agile leopard makes less bulky animals its prey, often

Red uakaris live in small bands high in the canopy of the Amazon forest, rarely descending to the ground. When threatened, they shake the branches, and their faces, which are normally very highly colored, turn an even brighter and more astonishing red.

climbing in order to do so, in South America there is no such division of labor. The jaguar has to fill both of these roles as best it can. It is hardly surprising that the jaguar looks like a compromise, having the leopard's relatively small size, but the tiger's heavier build.

In open clearings a smaller, long-bodied cat, the jaguarondi, matches the speed of the large South American rodents on which it feeds. Hunting only in the more open places, the jaguarondi has a virtually plain colored coat to blend with its surroundings. By contrast, the thicker parts of the forests are the haunts of three species of cats that, like the jaguar, have handsome blotched coats to provide camouflage. These are the ocelot, margay, and American tiger cat. The three differ slightly in size, and there must surely be important differences between them in the ways that they live and hunt that enable them to live side by side in the forest. However,

these differences still remain to be discovered, for here, as in so many ways, our knowledge of the animals of the tropical forests of America is very incomplete.

In the tropical forests of Africa the air-temperature is high and very steady, and insects, other invertebrates, and small vertebrates are abundant. These conditions of warmth and plentiful food are ideal for reptiles. On the ground live heavily built snakes with bold blotched markings that break up their outline and camouflage them among the fallen leaves and low branches. One of the best camouflaged and most deadly of these is the Gaboon viper, which may have ready for instant use enough venom to kill 12 men.

In the trees are slender, more lightly built snakes, able to stretch easily from branch to branch, and usually brown or green in color. When these plain-colored snakes keep still they

The tamandua is an anteater found in thick forests from Mexico to Peru, and its prehensile tail helps in climbing trees. Its long sticky tongue, too, is dextrous, seeking rapidly through the galleries of a broken termites' nest or pushing into bark fissures in search of the insects that go to make its diet.

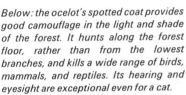

Below: the ocelot's spotted coat provides good camouflage in the light and shade of the forest. It hunts along the forest floor, rather than from the lowest branches, and kills a wide range of birds, mammals, and reptiles. Its hearing and eyesight are exceptional even for a cat.

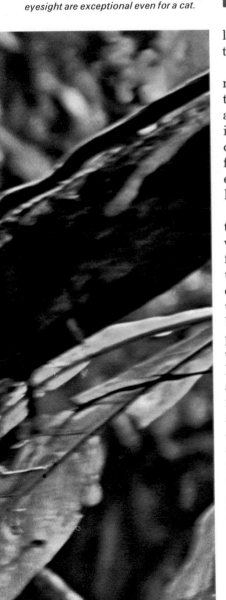

look either like the branches themselves or like the green stems of climbing plants.

Like other reptiles, snakes can remain still for many hours on end. They remain in a kind of trance, but they show a rapid response when alerted, like burglar alarms or mantraps, waiting inactive and apparently lifeless until triggered off. It is difficult to say whether they are asleep, for snakes' eyelids are transparent, and their eyes always appear to be open. In fact, the eyelids are fixed and they cover the eye permanently.

Against the changing patterns of light filtering through the trees, the chameleons, which are widespread and numerous in Africa's tropical forests, are well camouflaged. They have an unusual advantage in that their skin color changes to match their background. Although they climb only slowly, they are of all lizards the best adapted to life in the trees for they have both prehensile tails and strong, gripping toes. While they cling, almost invisible, they keep a sharp look-out. Their eyes swivel rather like the guns of a warship swiveling in turrets, so they can look in all directions without the head itself moving at all. Each eye moves quite independently of the other. If an insect comes to rest nearby, the chameleon eyes it, as if assessing the range, and then shoots out its incredibly long, sticky-tipped tongue. In a split second the insect has disappeared, and only the movement of the chameleon's jaws gives any clue as to where it has gone.

High in the forest canopy live many species of

The chameleon is among the slowest crawling reptiles of all, but compensates by its ability to match its skin color with the surroundings. It snipes unwary insects with a flick of its tongue.

birds, the brilliant hues of their plumage flashing in shafts of sunlight that pierce the gloom. Even the most brightly colored of these birds, such as the orange-breasted parrot and the small, stout-bodied lovebirds, tend to be inconspicuous in the dim twilight characteristic of Africa's rain forests. Most of them feed on the seeds and fruits of the trees, but the long-billed, iridescent sun-birds have a life style similar to the humming-birds of other forests, feeding on the nectar of the forests' blossoms.

Fiercest of the predatory birds of the African forests is the crowned hawk eagle. In most eagles the talons are the most powerful weapons, the hooked bill being of lesser importance until the prey is dead and ready to be ripped to pieces, but the crowned hawk eagle has a bill as strong and formidable as its talons. No eagle is more deadly. It can easily kill large hornbills, and sometimes snatches young monkeys from the treetops.

Probably the most remarkable life style of all the birds of Africa's tropical forests is that of the indicator bird or greater honeyguide. The favorite food of the adult indicator bird is wax from the honeycombs of wild bees' nests, but on its own it is incapable of dealing with a nest of angry bees. To solve this problem the bird has an incredible pattern of behavior that looks highly intelligent but is, in fact, purely instinctive.

On finding a bees' nest the bird chatters loud and long until it attracts the attention of a ratel, or honey-badger, a heavily-built carnivorous mammal. Moving in front of the ratel, and still calling, the bird leads the mammal to the bees'

The basic structure of the tropical forest (African in this case) consists of an understory at around 30 feet, a canopy at 120 feet, and above this, an emergent layer of scattered forest giants.

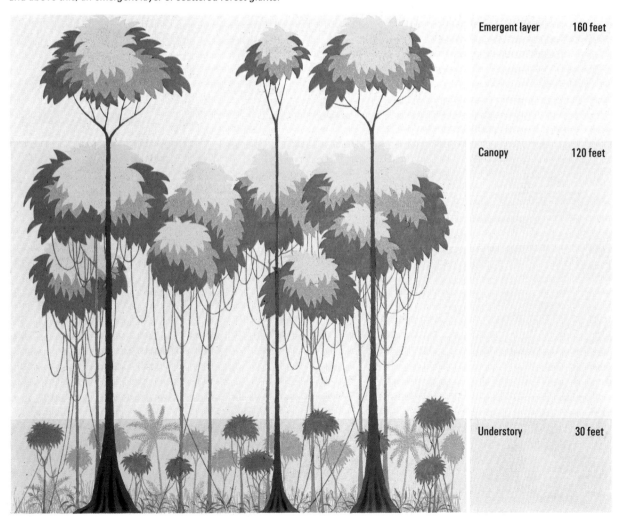

Emergent layer   160 feet

Canopy   120 feet

Understory   30 feet

The Levels of a Tropical Forest

Turacos are noisy, fruit-eating birds of the African forests; the species shown above is Donaldson's turaco. Members of this family have a number of interesting peculiarities, including outer toes that can be turned forward or back, according to the bird's immediate needs. Some species have a red pigment that is unique in the animal kingdom because it is water-soluble and may be washed out by the rain.

Right: the crowned hawk eagle is often seen soaring over the canopy of the Ituri forest in Zaire. It searches relentlessly for young monkeys, which are its chief diet.

nest, and waits while the ratel deals with the nest and feeds on the honey and bee grubs. Once the mammal has fed, the indicator bird moves in to take the wax comb left in the open nest as its own share.

This is a remarkable example of animals of different species behaving in concert in ways that are to the benefit of both. A relationship of this kind is called *symbiosis*, which comes from Greek words for "living together." In the absence of a ratel the indicator bird has been known to lead men to bees' nests. As long as the nest is opened and the bees dispersed, it is not important to the bird what animal does this.

The ratel, although fond of honey, also feeds on small mammals, reptiles, and insects. Its striking black and white markings make it very conspicuous, but being bold and powerful, it has little to fear from enemies. Beyond this, like the skunks of America, it can discharge a vile-smelling fluid from glands beneath the tail and so its black and white coloring serves as a warning to predators.

Scurrying along small, well-concealed paths in the undergrowth are big-eyed elephant shrews searching for food such as ants and termites. These little animals often make burrows for hiding in, but they may retreat into crevices among rocks, or under fallen trees for safety. Although

they probably rely largely on their sense of smell for finding food, elephant shrews are also said to stamp on the ground with their hind feet, making a noise to which the insects respond so that the elephant shrew is able to locate them.

Other eaters of small, social insects are the forest-living anteaters, or pangolins. They have no teeth but their tongues are very long and sticky. Because of this it was once thought that they were related to the edentate or toothless anteaters of America, but this is now realized to be another case of convergent evolution. The pangolins are covered by large overlapping scales that look rather like a cluster of dead

leaves, and act not only as armor but also as camouflage as the animals climb among the trees, clinging with their prehensile tails.

By contrast to the pangolins, the rabbit-sized tree hyraxes have short tails, and climb by sheer agility. The flat soles of their feet are always moist, and this helps them to move sure-footedly when at night, grunting and shrieking at one another, they climb in small groups, hunting for insects and succulent shoots on which to feed.

While the tree hyraxes are awake the big fruit bats leave their day-time roosts among the branches. Unlike their smaller, insect-eating relatives, they have large eyes, and find their food by sight and smell. They feed mainly on fruit and occasionally on flowers, taking the food into their mouths, crushing it and swallowing the juices before spitting out the pith and pips. In this way, they play an important part in spreading the seeds of some of the trees of the forest.

Under the cover of darkness many other animals go about their business in the African forests. Unlike South America, Africa has no nocturnal monkeys, but other, distantly related animals fill similar niches. The potto, for example,

Honeyguides have an uncanny skill at locating bees' nests. They flutter and call until a honey-badger or man comes and takes the honey, and they then eat the wax. They can also eat insects.

The zorilla is Africa's counterpart of the skunk. Though not very closely related, the two animals are remarkably similar.

may be seen climbing slowly through the trees at night, searching for fruit, insects, and small birds. Only when pouncing upon animal prey does it move with any sign of haste. It is well adapted to life in the trees for it has huge hands and feet, and on its hands, the forefinger is reduced to a mere stump. Because of this, the potto can cling to quite large branches, as the span of its hand stretches between the thumb and middle finger. The tail is only short, for, as the potto is small and always keeps a tight hold, balancing presents no problems.

While sleeping during the day, curled up and holding on to a branch with all four limbs, the potto has little to fear from most would-be predators. If disturbed, it straightens its legs and hits the enemy in the face with the long spines that project backward from the shoulder region of its backbone. Although these spines are fur-covered, their sharpness can still be felt through the animal's skin and they form an effective "knuckle-duster."

Pottos are found in most African forests but the closely related angwantibo, or golden potto, inhabits only forests of West Africa. Where the two species overlap there is room for both, since the angwantibo is smaller and lighter, weighing less than one pound, and can climb on thinner branches to reach food inaccessible to the potto.

Also active at night are the acrobatic bushbabies or galagos. While the pottos move slowly along clinging firmly to the large branches of the trees, the bushbabies use their long hind legs to leap swiftly from one branch to the next, sometimes calling loudly to one another. In the dim light they use their large movable ears as well as their eyes to find their way around.

Although they are sometimes said to be solitary, bushbabies often huddle together in quite large groups when they sleep in hollow trees during the daytime. Like their distant cousins the monkeys, they spend a considerable amount of time grooming each other, using their lower front teeth as combs.

By day, troops of monkeys take over from the bushbabies as the most agile climbers. Surprisingly, Africa has no precise counterparts to the marmosets and smaller monkeys of the tropical forests of South America. The talapoin is the smallest African monkey and it has a very restricted range, being found in only a few areas of rather swampy forest. However, in every other respect Africa has monkeys in a variety at least as rich as that of any other continent.

With their keen eyes positioned on the front of their heads, monkeys can judge distance well, an important faculty when leaping from one tree to the next. Their long fingers and toes can easily grasp large branches, each finger having near its tip a touch-pad, finely ridged to improve its gripping qualities, and supported by a flattened claw, the nail. Typical of the best climbers are the long-tailed and fierce-looking mangabeys, which usually live high in the forest canopy. Although large, they are built like athletes, lean and with powerfully muscled limbs. Like most monkeys, they are omnivorous, but are particularly fond of fruit, and range through the forest in troops, searching out trees where the fruit is ripe.

The guenons are a little smaller, and much more elegant in appearance. There are several species, some of which live exclusively in certain areas of Africa, while others share the same forests, but keep to different layers of the canopy. For example, the green monkey inhabits forest fringes and sometimes the edges of grasslands beyond, while Dent's monkey and the owl-faced monkey inhabit the same remote forests near the headwaters of the Zaire, or Congo River, but have slightly different life styles, and therefore are not in direct competition with each other.

Like the American marmosets, each different kind of guenon has distinctive markings and colors, especially on the face and the rump, and

The honey-badger, or ratel, is found throughout most of Africa and Southeast Asia, where it feeds on insects, reptiles, and even young antelopes. Above all, though, it loves honey, and often works in partnership with the honeyguide to find this favorite food.

Right: the giant pangolin lives in the more humid parts of the forests of equatorial Africa and may reach 6 feet in length. Its tiny head houses a long, agile tongue that reaches deep into termite nests and insect-infested trees.

Below: elephant shrews take their name from their long, mobile snouts. Different species are found in every African habitat, from arid bush to humid forest. This is one of the forest-dwelling kinds, here seen eating a cricket.

these help it recognize members of its own species. They live in large troops and, at intervals during the day when they are not feeding, groom each other, using their long fingers to go through each other's coats. They appear to find this activity very soothing, and it not only ensures that they keep clean, but also promotes sociability, helping to unite the troop.

The colobus monkeys, with their long, silky, and strikingly patterned fur, are a noticeable element of the animal life in the treetops. Hurling themselves from one tree to another, and landing in the branches with a resounding crash, they appear not to be handicapped by the fact that they lack thumbs, for they climb superbly. They feed mainly on leaves, which they grip between their fingers and the palm of the hand.

In forests near the Gulf of Guinea drills and mandrills live on the ground and in the lowest branches. They are rather heavy with powerful jutting jaws, and in keeping with their largely ground-dwelling existence their tails have become considerably shortened. They live in small troops and feed on fallen fruits, roots, fungi, and small animals. The dark-skinned drill usually keeps to the cover of the deep forests, and the larger, brightly colored mandrill more often lives in and near clearings.

In contrast to the drills and mandrills, the tailless chimpanzees climb well, often using their long arms to swing hand over hand beneath the branches. In forests to the north of the Zaire River lives the common chimpanzee, while the smaller, rarer, and dark-faced pygmy chimpanzee, or bonobo, lives to the south of the river. Since neither of these chimpanzees can swim,

rivers are impassable to them and the Zaire River forms an effective barrier.

Chimpanzees live in large troops and tend to keep to the same broad area of forest. However, this area is not a territory in the technical sense, for they do not defend it against neighboring groups of chimps. When two such groups meet they react by shrieking excitedly, but this appears not to be a threat display, for the groups sometimes join forces for a period.

In recent years chimpanzees have been the subject of much study, for as intelligent animals and probably our closest nonhuman relatives, they are of great interest to man. There is considerable disagreement about the details of the life of wild chimpanzees, as for instance, how much time they spend in the trees and how much they are on the ground, and whether they regu-

larly kill and eat such animals as small antelopes. Perhaps these differences of opinion are to be expected for, since chimpanzees are so intelligent, their behavior depends less on instinct and more on learning than that of other animals. The way in which a chimpanzee behaves probably depends upon the culture of the groups into which it is born, in very much the same way that human behavior does.

Living in rather smaller groups than the chimpanzees are the closely related gorillas. These great apes are shy, and are found only in dense forests near the Gulf of Guinea, and the thinner but inaccessible mountain forests of central Africa. Adult male gorillas rarely climb very far above the ground, for they weigh around 500 pounds. Females and young ones climb rather more, but the troop usually moves along on the ground, walking on all fours, on the knuckles of the long fore-limbs, and the flat hind feet.

Although the only animal to stand any chance of overpowering a gorilla is the leopard, the troop usually moves inoffensively away from trouble, only turning to fight if forced to do so. They feed mainly on plants, being particularly fond of some juicy stems similar to those of celery. In order to reach the soft and edible pith in the middle of harder plant stems, gorillas use their formidable jaws and teeth. The size of a gorilla's head is less due to the size of its brain—only about half the size of a man's on average—than to the huge jaw muscles, which are fixed to a ridge on top of the skull.

For the hoofed animals of the African forests life is more solitary and there is less inclination toward banding together in large groups than there is among the chimpanzees and gorillas. Among the towering trunks of the forest trees, festooned with the stems of climbing plants it is impossible to see far, and there would be no advantage in living in large herds. At the most, the hoofed animals of the forests, usually smaller in size than their grassland relatives, live in small groups. To locate possible danger they use their keen sense of smell, and keep their large ears swiveling, listening for the faint sounds that may betray an attacker.

Hiding beneath overhanging branches the shy little forest duiker is constantly alert and ready to dive through the bushes in order to escape from danger. Being only about 15 inches tall this antelope is well adapted for quick movement through the undergrowth. The horns, which are borne by both sexes, are short and so do not become entangled in the overhead branches. Even smaller than the duiker is the royal antelope, which is only 10 inches tall, and is found in the thick forests of West Africa.

In addition to these small, light-footed animals there are others, less agile, which make their

Left: the potto is widely distributed in Africa's tropical forests, where it lives at understory level. It has three spines at the base of its neck, which it uses in self-defense. Its powerful hands, with their large span, are ideal for clinging to branches, as the baby potto in the picture above demonstrates.

Right: bushbabies are born at all times of the year. They grow rapidly and soon become independent. Because of their mobile ears and huge-fronted eyes, bushbabies are particularly sensitive to their surroundings, and are able to find their way in the dim light. Although they belong to the same family as the potto, they are far more acrobatic, making spectacular leaps among the trees.

home in the African forests. For instance, African elephants of a special, slightly smaller forest-dwelling race crash their way through the branches, and bush pigs, and, in mountain forests of Central Africa, giant forest hogs, root in the forest floor.

Deep in the heart of the thickest Zaire forests the okapi, a member of the giraffe family, keeps to well-trodden paths usually traveling singly or in small family groups. So timid and well camouflaged are these beautifully marked animals that although they are large, standing about five feet tall at the shoulder, it was not until the beginning of this century that they were discovered by Europeans. Browsing on the bark, buds and leaves of forest trees and bushes, okapis use their long tongues to grip branches and pull them toward their mouths. A measure of the length and flexibility of the okapi's tongue is the fact that an okapi can lick its own eyes in order to clean them thoroughly

Above: a dwarf bushbaby, or galago. This animal is so small that it would easily fit into the palm of a man's hand.

Illustrations at the top of these pages show two main methods used by primates for getting from tree to tree. The indris, above, leaps from branch to branch high among the treetops of the mountain forests of eastern Madagascar.

A vervet monkey nursing her baby. Vervets are one species of the widely distributed grass monkey of Africa. They travel about in large troops, foraging in grass but rarely moving far from trees.

Apart from man, the leopard is the main enemy of the okapi, the chimpanzee, and the other large mammals of the African forests. Like almost all cats, leopards usually hunt alone, approaching their prey stealthily, and using the considerable muscle-power of the body and hind legs to pounce from a short distance. The large canine teeth and sharp claws are formidable weapons and a 100 pound leopard can easily rip open the throat of any forest animal except an elephant. To ward off this attack the animals of the forest have as defenses their own watchfulness, camouflage, and speed in escaping. The struggle for survival is evenly matched.

Leopards usually hunt in the evening or at night. In the failing light the dark rosettes on the leopard's coat blend with the forest shadows. Approaching its prey the leopard moves almost soundlessly on its soft pads, for its claws are retracted. They are only pushed out when they are needed to grip the bark of a tree, or for use as weapons. Leopards climb well, using their long tails to balance themselves, and often approach their prey through the branches, and dive from above. Sometimes they lie in ambush above a forest path, and attack any large animal that comes along. Even men, greatly feared by almost all wild animals, are sometimes attacked in this way. The leopard rather than the lion, which is usually found in grasslands, is the king of the jungle.

Apart from leopards there are surprisingly few cats in the African forests. The rare African golden cat lives in some West African forests, and in thinner forests and bush country the African wild cat, a close relative of the European

The gibbon swings from arm to arm, a method characteristic of the anthropoid apes. Often they can get up enough speed in this way to carry themselves over 20-foot gaps. Even a youngster clinging to its mother's waist seems to be no impediment.

Colobus monkeys are the long-jump champions of the high treetops, often crossing gaps so wide that the leap would appear to be an impossible feat.

wild cat, is not uncommon. However, Africa lacks the range of smaller forest-dwelling species of cats to be found in either South America or southern Asia.

Unlike America, Africa has resident members of the mongoose family, and perhaps these fill the gap left by the absence of small cats. Certainly some of the mongoose family are very catlike in appearance. The little genets, for example, have blotched catlike coats, long tails, and quite long legs armed with claws that can be partially retracted; only their slightly longer noses distinguish them easily from cats. Genets are nocturnal, and hunt both in the trees and on the ground, killing small mammals, birds, and reptiles, and also feeding on fruits.

The most numerous members of the mongoose family are the mongooses themselves. They, too, are omnivorous, and include among the animal food that they eat insects, reptiles, birds' eggs, and mammals up to the size of a hyrax.

Many of the mongooses live exclusively in the forests on the huge island of Madagascar, which is the home of many unique animals because it lies isolated off the coast of East Africa. Hunting in these forests during the day, the ringtailed mongoose, for instance, is a good climber but it flees to the ground if surprised in the trees. Unfortunately, there must be considerable doubt as to whether many of these island species of mongoose will survive much longer because parts of Madagascar have been deforested by man in recent years.

Living exclusively in the rain forests that lie toward Madagascar's east coast are some species of the tenrec family, which is unique to the island. Some of these small, short-legged, thickset mammals have long tails and are well adapted

Right: chimpanzees never move far from trees, although they spend much of their waking day upon the ground. There they search for nuts, berries, insects, and fruit. Occasionally they will kill small antelopes, which the troop leader will share out with great ceremony.

Gorillas are the largest and heaviest of the primates. They live in family groups, roaming the forest floor during the day, foraging for leaves, stems, and other plant material on which to feed. At night they sleep in nests, which they build either on the ground or in the trees. Above: a lowland gorilla. These fairly common gorillas live in pockets of forest between the Zaire and Niger rivers. Left: a group of mountain gorillas. These are comparatively rare and very shy animals, which live peacefully in the remote forests bordering the Rift Valley between Zaire and Tanzania.

to life in the trees. Others, such as the spiny tenrecs, are more suited to life on the ground for they lack long tails. They are well protected against predators, having armor composed of stiff, bristly spines all over their backs. All tenrecs have sharply pointed teeth, which they use to feed on worms, snails, insects, and some fruit.

Climbing clumsily through the branches the aye-aye, another of Madagascar's unique forest mammals, listens for the faint sounds that betray the presence of beetle grubs under the bark. On detecting these sounds it uses its large front teeth to pierce a hole in the bark and then hooks out the grubs using its remarkably long third finger. Being omnivorous, the aye-aye also feeds on fruits such as unripe coconuts. It chisels off the top of a green nut using its teeth, and then puts the end of its long finger into the coconut milk, and the base of the digit into the corner of its mouth. By moving the finger rapidly to and fro the milk is taken to its mouth a little at a time, clinging to the surface of the finger.

Benefiting from the lack of competition resulting from Madagascar's long isolation are the lemurs which, like the aye-aye, are distant, tree-dwelling relatives of the monkeys. These mammals climb by the same method as bushbabies, kicking off powerfully with the long hind legs, flying through the air, and seizing the target branch with all four limbs. They are sociable animals usually living in groups, and their

Long legs and a long neck transform the slender gerenuk into a 6-foot browser of otherwise unreachable leaves. Gerenuk live in pairs or small herds in the East African bush, and never venture out into the open grassland away from the leaves that form their food.

systems of communication are well developed. Their varied calls, which may sound like the yowling of cats or the grunting of hogs, add to the "voice" of Madagascar's tropical forests.

The largest of the lemurs, such as the ruffed lemur and the ring-tailed lemur, are active by day. Their bodies are handsomely patterned in a wide range of colors and markings, for their eyesight is good and, like man and birds, they have some color vision. These day-active species communicate by visual signals. The ring-tailed lemur also uses scent signals. Scent produced by glands on the wrists is rubbed on the prominent black and white striped tail, and when the lemur curls its tail forward, and then flicks it back over its shoulder and waves it, the combined action

of scent and sight stimulates other lemurs.

Other smaller and more dully colored lemurs are nocturnal. These include the lesser mouse lemur, which is only a little larger than a mouse. Like other lemurs, it is omnivorous and especially fond of fruit. Because it lives in forests subject to seasonal rains, which affect the fruiting of the trees, the lesser mouse lemur has become adapted to the problem of food shortage during the dry season. While food is plentiful it feeds greedily, becoming very fat and when the dry season comes, it makes a nest and goes to sleep for weeks on end. This is comparable to the hibernation (from Latin *hibernare*, to pass the winter) of animals in cold climates, but since it is a response to summer rather than winter con-

The okapi lives so deep in the Zaire forests that it was unknown until 1901.

A rare picture of the bongo, which lives perfectly camouflaged in the deep, equatorial forests of Africa, from Zaire to the Aberdare Mountains in Kenya.

The masked palm civet is found from China to Borneo. A superb climber, it thrives on fruit, insects, small vertebrates, worms, and grubs. It is also skilled at catching fish.

ditions, it is called estivation (from Latin *aestas*, summer).

Two of the most elusive of Madagascar's unique forest mammals are the rare sifaka and the larger, and equally rare indris. The diademed sifaka, one of several sifakas found in Madagascar, lives in the eastern rain forest, moving through the trees by jumping from one upright branch to another using its pincerlike back feet to grip. It is well adapted to life in the trees and it has a long tail.

In contrast to the sifaka, the indris has virtually no tail and is less obviously adapted to

The leopard is the most widely distributed cat in the world, being found in much of Africa and Asia. Like many cats, leopards are surefooted climbers, and will often drag the remains of their prey into the trees so that they can enjoy a leisurely meal out of reach of earth-bound scavengers.

climbing. Nevertheless it climbs well using its long limbs to leap from branch to branch. It lives high in the forest canopy, where it basks in the sun, for both sifakas and indrises are sun-lovers.

It is tempting to compare Madagascar's tailed lemurs and tailless indris with the monkeys and apes of the African mainland, and perhaps in some ways these groups do fill the same ecological niches. However, unlike the true apes, the indris does not swing by its arms, so the comparison must not be taken too far.

In the southern and Southeast Asian tropical rain forests live animals that are closely related to those of the African forests. For example, in both continents the rain forests are the home of similar species of cobras, and among the birds, the largest fruit-eating species belong to the hornbill family. The hornbill's long, strong bill is reminiscent of that of the toucans, and is similarly used to push through tangled foliage and reach fruits on the thinner twigs.

However, a unique assembly of flying—or at least gliding—animals is found among the amphibians and reptiles of these Asian rain forests. The flying frog glides from one tree to another,

These dwarf mongooses have appropriated an abandoned termite hill at the forest margin. Mongooses are agile hunters found throughout the warmer Old World regions. They live in large groups, and mark their territory with scent from their cheek glands.

121

The lesser mouse lemur is one of the smallest living primates and climbs on branches of all sizes, feeding on insects and fruits. By day it sleeps curled up in a ball, often with the tail twisted round a branch for support.

The indris, whose method of movement is shown on page 114, is called the "dog of the forest" by the people of its native Madagascar. Although it is, in fact, a leaf-eater, it has a doglike face, and a very loud, barking cry that echoes through the high treetops of the isolated mountain forests where it lives.

The aye-aye is a primate with the nails and teeth of a rodent. It feeds on the larvae of wood-boring insects, tapping the bark with its long nails to detect the presence of the grubs.

supported by its webbed feet. The paradise tree snake can spiral gently down from the branches, having flattened its body to form a "wing," and the flying lizard can glide for 60 feet on membranous "wings" supported by its ribs, which stick out on either side of the body.

Among the larger herbivorous animals of the tropical forests of Asia, deer are common. To a certain extent they replace the antelopes of the African forests, following the same rules with regard to body size and social habits. The larger deer with the longer antlers, such as the sambar deer, live in small herds in the more open forests, sometimes emerging to graze in the grasslands beyond, while the smaller deer are more solitary and lead furtive lives in the depths of the forests. Of the smaller species, the graceful spotted deer of Burma and India, and its relative the hog deer are still too tall to move easily through the thickest undergrowth. The smallest deer of all with short, inconspicuous antlers are the muntjacs or barking deer, which share the thickest forest cover with the wild boar, and with the even smaller chevrotains. Chevrotains are the survivors of an ancient group of hoofed animals and lack both horns and antlers, the males being armed instead with long, tusklike canine teeth.

Instead of moving silently underneath tangled branches as the smaller deer do, the heavy, horned cattle of the hilly forests of India, Burma, and Malaysia, known as gaurs, use their weight to crash through the undergrowth. Living in family groups, gaurs emerge from the shelter of the forests to graze in clearings, but in spite of their size they are shy animals, and retreat among the trees if disturbed.

The forests of southern Asia abound in low-lying swampy areas, and here the broad-horned wild water buffalo graze and wallow in the mud. They live in quite large herds and, instead of taking flight if a predator, such as a tiger, appears, they turn to face it and, if necessary use their combined strength to gore and trample it.

Also using bulk both to overcome the problems of getting through the thick forests and for self-defense is the Indian elephant. Found in Ceylon (Sri Lanka), Burma, Indo-China, Malaysia, Sumatra, and Borneo as well as in India, this species would be better named "Asiatic ele-

Right: like the mouse lemur, the sifaka (another lemur) is found only in Madagascar, but fossils prove that this group of mammals was once much more widely distributed.

phant." The species as a whole is much more forest-loving than is the African elephant.

Indian elephants live in small herds and feed at intervals throughout the day and night, with peaks of activity at dawn and dusk. During the heat of the day they wallow in the rivers in order to keep cool. The elephant's trunk, which is an elongated and well-muscled nose and upper lip, is an incredibly versatile organ. It is used for gathering food, for drinking (water being drawn into it and then squirted into the mouth), for dust bathing and bathing, as a snorkel when swimming in deep water, and as a weapon against enemies. With its great weight, long, prehensile trunk and, in the case of the male, its tusks, the adult Indian elephant is safe from all enemies except man, although tigers may sometimes kill calves if they can first succeed in distracting the attention of the mother.

Asian rhinos, too, use their bulk in self-defense. Although they are armed with horns, it is their teeth that do the most damage when they hurl themselves at an attacker, for, unlike the African rhinos, they have a pair of forward-pointing razor-sharp incisors in the lower jaw. All Asian rhinos have pointed, movable upper lips, which are used for tucking branches and leaves into the mouth. Like the elephants, the rhinos sometimes use their weight to push trees over in order to browse on the higher leaves and the exposed roots. Rhinos and elephants are also similar in that they both tend to wallow in mud in order to keep cool. Rhinos can be active at any hour of the day or night, but tend to become nocturnal if they are disturbed by man.

Living near water in the densest tropical forests of the Malay Peninsula and Sumatra are the timid Malayan tapirs. Like the closely related South American tapirs they are solitary, nocturnal animals and browse during the night on water plants and on the forests' lower leaves and branches. The Malay tapir is easily distinguished by the bold black and white markings which, in a moonlit forest, break up the outline of its body and serve as camouflage.

On the ground among the trees or the tall grass jungles of tropical Asia tigers hunt deer, buffaloes, and wild boar using the methods typical of the cat-family, stalking and pouncing from a short distance. The tiger is another animal whose markings help conceal its movements. Although it is the best-known of the large carnivores of southern Asia, it is a relatively recent arrival, for

The large "casque" on the hornbill's beak is an air-filled chamber that may help to modify the raucous cries the bird utters.

tigers originated in Central Asia, and some still live there. Only within the past few thousands of years have tigers invaded the tropics, and even today they are imperfectly adapted for this life, as they find great heat difficult to bear.

Climbing the trees and hunting slightly smaller prey are leopards, which are perhaps the most athletic of the cat family. Leopards hunt ceaselessly, their favorite prey being deer and monkeys, and they often drag the kill up into a tree so that the uneaten remains will be out of the reach of earth-bound scavengers.

As in the forests of South America, the Asian forests are the home of a number of species of smaller cats with spotted and striped camouflage. For example, the jungle cat and the fishing cat live in the swampy forests so common in Asia, and the leopard cat is found in forests of all

The green tree viper of the African forest freezes when it is threatened, its color providing a perfect camouflage. It preys on tree frogs and small mammals, taking a firm grip of a branch with its prehensile tail before it lunges and strikes its chosen victim.

kinds. The marbled cat and the very similar, but larger, clouded leopard have distinctively blotched, pale-centered markings, and long tails. They are thought to prey on roosting birds among the branches. Temminck's golden cat, found in thick forests from Tibet to Sumatra, looks very like the African golden cat, which is probably its closest relative.

The variety of cat species is even greater than in South America, and the reason for this is imperfectly understood. Some of these wild cats live exclusively in one or other part of the continent, but others apparently live side by side in the same forests. Although we know what the various species look like, in a very real sense we do not know them at all, for probably the most important thing to know about a species is where it fits among the complicated pattern of other

The dark-colored forest cobra, shown here, is one of the most venomous of all the reptiles found in Asia's forests.

Above: muntjac, or Indian barking deer, live in pairs, preferring dense undergrowth but often emerging into clearings on the forest edge. They have escaped from collections and established themselves in the wild in many countries.

Below: the Indian rhino is a shy and rarely aggressive animal, usually living alone in forest or reed swamps, where it adheres strictly to its own territory. Destruction of the rhinos' natural habitat in recent years, together with excessive hunting by man for their horns, has brought the species close to extinction.

Indian elephants are the largest animals in southern Asia. They browse in the forests and consume 30 to 50 gallons of water daily.

living things—what ecological niche it fills, and what are the reasons for its success in the unending competition between species.

The wild dog of the forests of tropical Asia, the dhole, is as much feared as the tiger for, like their distant cousins the wolves, the fierce dholes hunt in packs, running tirelessly in pursuit of their prey, especially deer.

The forests of Asia are also the home of the largest member of the mongoose family, the tree-dwelling, nocturnal, binturong. The binturong has long, shaggy, black hair and uses its prehensile tail like a fifth limb as it climbs in the trees, searching for fruit and other vegetation. It is comparable in some of its habits, and life style with the tropical American kinkajou. The related masked palm civet is equally at home in the trees, and like the binturong is nocturnal, feeding on fruit, insects, and small invertebrates.

Sharing a home in the branches of the forests of Burma, Malaya, and some of the neighboring islands with some of the world's largest and most colorful squirrels are the flying lemurs. Despite their name they are not primates, but the sole members of an isolated group of mammals with no close living relatives. Flying lemurs are about two feet long, but are very lightly built, and weigh only about three pounds. They have wings consisting of flaps of skin running from the throat to the hands, feet, and tail, and often glide for distances of 50 yards or more among tall trees, losing very little height in the process. Despite the encumbrance of their wings they also climb well. Like the fruit bats that they resemble in some ways, they feed on fruit, buds, and flowers, but unlike the bats, they swallow their food entire, rather than merely sucking the juice.

At first sight the little omnivorous tree-shrews that live in the undergrowth of the Asian forests look like squirrels, for they have the same long tails, and the same graceful yet jerky movements. However, unlike any rodent, they have long narrow heads and rather pointed noses. Good vision is important in climbing, and tree-shrews can see well with their large eyes set on the side of the head.

In contrast to the rapid, almost agitated movement of the tree-shrew, the lorises climb slowly and carefully in the trees. They rarely descend to the ground, spending the day curled up asleep in the trees, and waking at night to stalk silently through the branches hunting for small verte-brates and insects, and to forage for fruit. In their search for food, the lorises may come face to face with the big-eyed tarsiers that inhabit the tropical forests of Sumatra, Borneo and the Philippine Islands. Tarsiers look very much like bushbabies, being adapted for the same sort of life, leaping acrobatically from branch to branch by night.

Living high in the forest canopy are the langurs or leaf-eating monkeys, while on the ground and

Above: the binturong, or bear cat, often reaches 5 feet in length. It has an immensely long, shaggy coat and a big, bushy tail that accounts for about half its length. In spite of its ferocious appearance it is more of a vegetarian than a meat-eater.

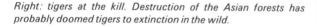

Right: tigers at the kill. Destruction of the Asian forests has probably doomed tigers to extinction in the wild.

in clearings the other major group of monkeys found in the Asian forests, the macaques, make their home. The macaques belong to the same group of monkeys as those of the Japanese forests. They are usually rather short-tailed with rather jutting jaws and they feed omnivorously. One species, which inhabits the Malaysian lowlands, is called the crab-eating macaque, for in the wide range of foods that it eats it includes both water-living crabs and land crabs.

The most acrobatic of the apes, the long-armed gibbons, swing through the trees in small family groups, defending their territory against neighbors by calling and hooting loudly—behavior that reminds us of the howler monkeys of South America. Again, there are several species, occurring from Assam to the islands of the East Indies.

The low-lying forests of Sumatra and Borneo are the home of the only great apes found out-

Orang-utans are the only great apes found in the wild outside Africa. Once they may have been common in Asia, but they now exist only in Borneo and Sumatra. They spend most of their time in trees, clambering about in search of fruit, leaves, and other food.

Above: the Philippine tarsier is a nocturnal primate that feeds mainly on insects. Its powerful legs can launch its 6-inch body on leaps up to 6 feet. Its feet have suction pads to aid in climbing.

side Africa, the orang-utans. They live in small groups, and feed mainly on fruit. They are particularly fond of the fruit of the durian tree, removing the spiny husk with their fingers and muscular lips, before eating the fruit. Although orang-utans can move quickly when they choose, they are not especially agile animals and they usually clamber somewhat lethargically in the trees, occasionally walking upright along a branch. At night they build "nests"—platforms of branches collected in the fork of a tree—but these are only temporary, perhaps being used for several nights before the group moves on.

With a high temperature and heavy rainfall to allow abundant plant growth the Australasian tropical forests are as luxuriant as any, but because of their geographical isolation they are populated by animals of types not found elsewhere. The reptiles of Australasia are most closely related to those of Southeast Asia, but have been separated from their relatives for so long that they have become very distinct and separate in type. For example, the most deadly

Above: the slow loris is a careful hunter, stalking small birds, geckos, and insects with great cunning. Its name comes from the Dutch loeris, *meaning clown.*

A female entellus langur rests in the fork of a tree. Some langurs are among the most highly colored primates, especially at the collar.

The shortnosed bandicoot of Australia excavates the forest floor in search of grubs and roots on which to feed.

Below: the flying fox, a fruit bat, has a wingspan of up to 5 feet. It is distributed widely in Australasia and Africa.

Below: the cuscus moves slowly in the trees. Its diet consists mainly of leaves and fruit, but includes small animals.

Below: Lumholtz's tree kangaroo uses its long tail as a balance as it browses confidently among the Queensland treetops.

of Australasia's many kinds of poisonous forest-dwelling snakes is the taipan, which can exceed 10 feet in length, and belongs to the cobra family. No human bitten by this snake is ever known to have survived. Fortunately the taipan is quite rare, mainly inhabiting the forests of the Cape York Peninsula.

New Guinea is one of the few places where the two main groups of large, constricting snakes, the boas and the pythons, can be found living side by side. There is a range of species, capable of killing and swallowing whole mammals and birds of varied sizes. The green python of New Guinea closely resembles the tropical American emerald boa in both appearance and behavior, lying in loose, leafy-looking coils among the branches, waiting for small vertebrates to pass by. The largest Australasian python, the amethystine python, can grow to be 20 feet long, and is able to overpower and swallow fully-grown wallabies. It is among the most formidable of all pythons.

Among the many remarkable birds of tropical Australasian forests are those of the closely related bowerbird and bird-of-paradise families. These birds are unique to Australasia, and both stand out for their incredible courtship displays. In birds-of-paradise it is the gaudy and often long plumes of the males that act as courtship signals. Bowerbirds are less remarkable in appearance, but the males build and decorate an elaborate bower to attract their mates. All the members of both families are forest-dwelling, and feed on fruits, seeds, and insects.

Equally remarkable and also found only in the tropical forests of Australasia are the three species of birds that make up the cassowary family. These are large, heavily built birds with small wings, and are unable to fly. They keep well hidden in the thick forest and if they are disturbed while feeding on the forest floor, they run to safety, kicking their way through the tangled vegetation with their powerful three-toed feet. If cornered they kick out at the attacker and can inflict severe injury, for their toes bear large, sharp claws.

In contrast to the giant-sized birds, the mammals of Australasia's tropical forests are all comparatively small, an unusually high proportion of them living in trees. As in the temperate forests of Australasia, the typical mammals of the tropical forests belong to only two groups, the egg-laying monotremes, and the pouched mar-supials, although some rodents and bats are present too.

The ground level of the tropical forests of north-eastern Australia is the home of egg-laying spiny ant-eaters of the same species that occurs elsewhere in Australia. The spiny ant-eaters of New Guinea are a little larger, and have even longer, beaklike, toothless snouts with long, sticky tongues that they use to feed on ants and termites. If frightened, they erect their spines and use their large claws to burrow rapidly out of the reach of enemies.

The mosaic-tailed rat of the forests of northern Australia and New Guinea, lives among branches where it feeds on leaves and berries. The young cling to the mother's breast for the first two weeks of life, and are dragged after her wherever she goes. Even when they are older, young mosaic-tailed rats cling to the mother if they are startled.

Apart from the dingos, the largest mammalian predator of Australasia's tropical forests is a marsupial, the northern dasyure. It is a nocturnal hunter, as big as a small domestic cat, and has brown fur spotted with white. The dasyure feeds on small vertebrates and invertebrates, and poses no serious threat to the medium-sized forest animals. It may be that in prehistoric times the larger Tasmanian devil (now confined to Tasmania) and still larger Tasmanian wolf or thylacine (now possibly extinct) were quite widely distributed, and filled the otherwise in-explicably vacant niche for larger carnivores.

The more open forests of tropical Australasia are the home of rabbit-sized, short-nosed bandi-coots, while the thick rain forests are occupied by the spiny bandicoots. The members of this family are nocturnal and usually hide in nests on the ground during the day. They feed on insects and other invertebrates.

As with herbivores in the forests of Africa and southern Asia, the forest-living herbivores of Australasia are rather small, and tend to be solitary. For example, the little scrub wallabies or pademelon wallabies, which are members of the kangaroo family, are less than three feet tall. Keeping to tunnel-like paths through the undergrowth, they feed on leaves and shoots, which they pull toward their mouths with their forepaws.

The largest climbers of the tropical forests of Australasia belong to the kangaroo family, too. They are the tree kangaroos. Although they have

Above: the red-plumed bird of paradise, like other such birds, is found only in and near New Guinea. Bright color well displayed helps attract mates in the dim light of the forest.

The cassowary has a bony casque, and wings with unfeathered quills, so it is able to run through undergrowth without becoming entangled. The male bird incubates the eggs and rears the young.

shorter hind limbs than other kangaroos, they are heavily built, and show little sign of being well-adapted for tree climbing. Their powerful claws give them a grip on the bark, and their long tails enable them to balance when hopping from one branch to another, or from tree to ground. It is said they can jump from branches over 60 feet high to the ground without hurting themselves. Nevertheless, they are not truly expert climbers, for when they climb down a tree-trunk they descend tail first, rather than head first as the very best of climbers do. However, tree kangaroos do not need to be better climbers in order to survive, for they are, in fact, the best of their size to be found in any Australasian forests.

Of the other tree-living marsupials that live in

The amethystine python of New Guinea is well adapted for life among the trees, with its excellent camouflage, and its prehensile tail for gaining leverage.

the tropical forests of northern Australia and New Guinea some resemble dormice, while others look like squirrels. The most unusual in appearance are the nocturnal cuscuses, which look rather like lemurs with thick, prehensile tails. At night these animals climb slowly through the trees at a leisurely pace reminiscent of that of the slow loris of the Asian tropical forests. Their stealth enables them to surprise small mammals and roosting birds with which they supplement their staple diet of leaves and fruit. If they are handled, cuscuses emit a strong musky smell from scent glands, but it is doubtful if this is a very effective defense or, indeed, if the cuscuses even need to defend themselves. Apart from large pythons they have no enemies to fear.

Smaller and more lightly built tree-climbers in

The diagram overleaf:

This tropical forest food web is based on the Amazonian forest of South America. Such forests are characterized by a vast range of tree species of varying form. Some, such as the kapok tree shown in the illustration, tower above the forest at intervals. Most of the remainder form a continuous evergreen canopy at a lower level. Creepers and epiphytes such as the bromeliads thrive throughout the forest. The climate in these regions is one of high temperatures and rainfall throughout the year. This lack of seasonality allows continuous flowering and fruiting within the canopy, making more opportunities for animals to specialize on very restricted diets. For example, hummingbirds, honeycreepers, toucans, and fruit bats all depend upon a steady supply of flowers or fruit. Similarly, jacamars depend upon wasps or bees, which in turn depend upon continuous flowering to remain active all year.

Apart from this, the considerable structural diversity and range of plant types tend to increase the number of specialist niches available. For example, the primitive hoatzin is especially adapted to feed on only Arum leaves. These specialist niches, which are characteristic of all tropical forests, are generally similar whether in Southeast Asia, Africa, or South America.

135

A Simple Food Web in a Tropical Forest

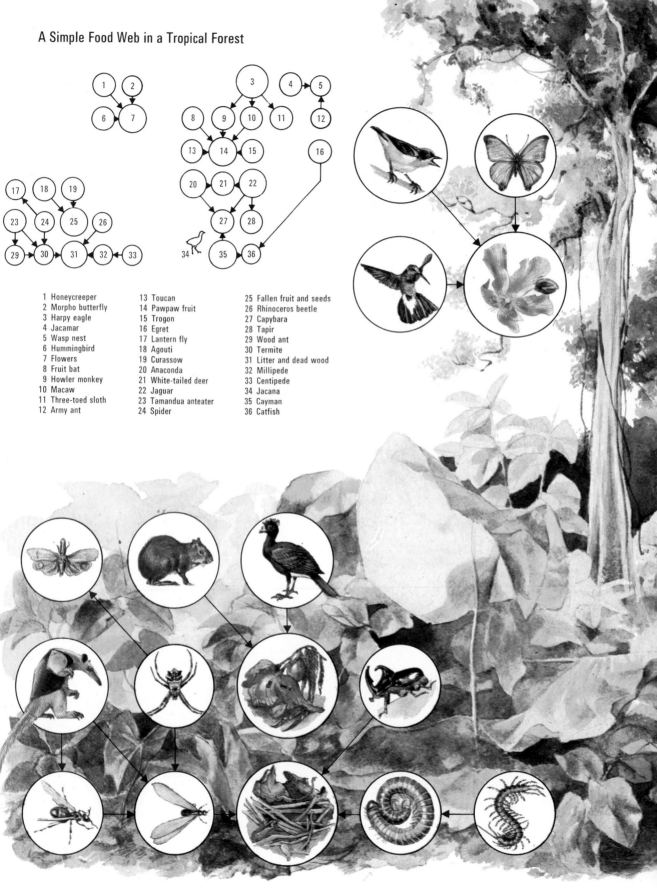

1 Honeycreeper
2 Morpho butterfly
3 Harpy eagle
4 Jacamar
5 Wasp nest
6 Hummingbird
7 Flowers
8 Fruit bat
9 Howler monkey
10 Macaw
11 Three-toed sloth
12 Army ant

13 Toucan
14 Pawpaw fruit
15 Trogon
16 Egret
17 Lantern fly
18 Agouti
19 Curassow
20 Anaconda
21 White-tailed deer
22 Jaguar
23 Tamandua anteater
24 Spider

25 Fallen fruit and seeds
26 Rhinoceros beetle
27 Capybara
28 Tapir
29 Wood ant
30 Termite
31 Litter and dead wood
32 Millipede
33 Centipede
34 Jacana
35 Cayman
36 Catfish

the Australasian tropical forests include the tiny pygmy possums, the handsome and bushy-tailed striped possums, and the smallest of the gliding marsupials, the pygmy gliding possum. All of these climb more rapidly than the cuscuses and can run down tree-trunks head first.

Sharing the treetops with these marsupials are climbing rodents, such as giant naked-tailed rats, which feed on coconuts, and rabbit rats, which hide by day in hollow trees, emerging at night to feed on the fruits of the corkscrew palm.

In the East Indian islands the animals of Asia and of Australasia sometimes meet and blend to form a community of unequaled variety. It is a transitional zone where the fringes of two different populations of animals overlap, and where many of the species are unique, having been cut off from their closest relatives by the sea for millions of years. For example, in the tropical forests of the strangely shaped island of Celebes, Australasian birds and marsupials live alongside short-tailed and long-haired monkeys of species not found elsewhere. Big-eyed and bushbaby-like tarsiers leap through the trees, while the thick forest undergrowth conceals the smallest of the forest-dwelling buffaloes, the anoa, and also a unique and long-tusked pig, the babirusa.

Like all of the world's forests, those of the East Indies are seriously threatened. The human population explosion amounts to a plague of men, quite recent in origin, that threatens the existence of all other life-forms and living communities. By felling the trees for timber and making clearings for at first agriculture, and then freeways, suburbs and towns, we have destroyed and are still destroying not just a wilderness, but a priceless part of our heritage, and a vital one, too. The very oxygen that we breathe is produced by living plants as a by-product of the process by which they build up their own food materials. If we destroy all of the forests there is a danger that the composition of the earth's atmosphere, so important to us, could become permanently affected.

As we have seen, the plants and animals of the forests do not live in isolation but each depends on the other. So, too, do the varied plant and animal communities that make up the world that we know. The balance is easily upset. If one kind of living thing in a community is destroyed, the effect can be widespread, and if one type of community is entirely lost, there must be serious effects on those that still remain.

The forests that men sometimes create are not the same as those we are in danger of losing entirely. Tract forestry produces a checkerboard of trees of a single species and age, stretching as far as the eye can see, and broken only by regularly spaced firelanes. This is a poor substitute for the diverse wealth of natural forests, which are the most luxuriant and abundant natural communities that the world has ever seen. If they should be lost it will cast a strange light on the meaning of the word "progress."

In the long perspective of geological time spanning hundreds of millions of years, forests

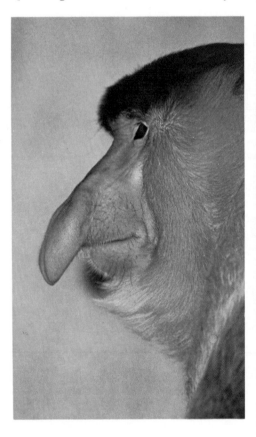

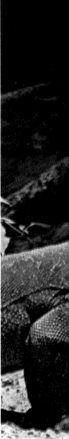

Above: the proboscis monkey. This monkey is found only in the forests of Borneo, and is so-called because the male of the species has a large, bulbous nose, which may be as much as 3 inches in length. The function of this remarkable structure is still not known for certain.

have been shaped and perfected by slow but unrelenting forces of evolution. Each of their myriad species of plants and animals has become adapted to fill its present place within the community. Those that now survive are the most perfectly adapted of all, for many species have lost out in the struggle for existence, and become extinct. Only the best remain.

Man is a relatively recent arrival on the scene, having existed as a species for only one or two million years. For most of human history the forests have seemed inexhaustible and unending. The same wild species lived out the same roles, and the same chemical substances were recycled again and again. Individual plants and animals died, but were replaced by others of their own kind. All that was used up was the energy used by living organisms in growing, feeding, breeding, hunting, and escaping, and this energy was endlessly replenished by the sun. The system had its own momentum and balance, and its own apparent permanence, like a spinning whip-top. It is terrible to contemplate the possibility that, for lack of wisdom on the part of man, the cleverest of animals, this intricate natural beauty may be lost.

The Komodo "dragon" is the world's largest lizard. Found on Komodo and nearby islands of Indonesia it averages between 6 and 9 feet in length and is both powerful and fierce enough to attack and kill quite large mammals, such as this goat.

Index

Picture Credits

144